T0223511

The Wonders of Marine Plankton

Albert Calbet

The Wonders
of Marine Plankton

 Springer

Albert Calbet
Institute of Marine Sciences, CSIC
Barcelona, Spain

ISBN 978-3-031-50765-6   ISBN 978-3-031-50766-3   (eBook)
https://doi.org/10.1007/978-3-031-50766-3

Translation from the Spanish language edition: El plancton y las redes tróficas marinas by Albert Calbet, © The Author 2022. Published by Los Libros de La Catarata and CSIC. All Rights Reserved.

This Springer imprint is published by the registered company Springer Nature Switzerland AG
The registered company address is: Gewerbestrasse 11, 6330 Cham, Switzerland

Paper in this product is recyclable

To all those who recognize the preciousness and beauty of the ocean and its inhabitants, and who believe in their preservation and protection.

Introduction

I would lie if I said that I have always been passionate about the sea. As a child, I experienced the motion sickness caused by boats, which was far from pleasant for my stomach. However, it is true that since my childhood, I have been fascinated by aquatic microorganisms and living beings in general. I suppose everything changed for me on December 31, 1976 (the day I turned 8), when my parents gave me a toy microscope. That simple plastic device with low optical quality determined my destiny.

I still remember my first unicellular organisms, collected from water in a sun-exposed flower vase. The tiny ciliates and green algae that filled my retina left an everlasting mark on me. That day, I decided I would become a biologist and study microorganisms. As the years passed, fueled by that idea, I used my first paycheck to buy a new microscope, to which I attached a rudimentary camera. Then another one, almost professional, with which I took numerous microphotographs. I began my studies in Biology and continued pursuing my passion; no puddle, reservoir, or container of putrefying water could resist my curiosity. Gradually, I started my own cultures of protozoa and algae, improving my photographic technique to the point of participating in a photography exhibition at the Faculty of Biology in Barcelona. It was those photographs that caught the attention of some teachers and paved the way for me to receive a scholarship to carry out a doctoral thesis on marine zooplankton at the

Institute of Marine Sciences, CSIC, under the supervision of Professor Miquel Alcaraz. And that is where it all began.

After three decades dedicated to the study of marine plankton, I believe the time has come to share my knowledge of this fascinating world, a privilege that only a few of us have had. While it is true that most people are aware of jellyfish, which can prove bothersome during summer, and have heard about the detrimental impact of plastic pollution in our seas, the warming of our oceans, and the acidification that harms coral colonies worldwide, there is still much more to comprehend. Few individuals truly grasp the significance of plankton in our lives, their amazing beauty and intricate complexity, and, most importantly, the urgent imperative to conserve them. Perhaps a handful may recognize terms such as protists, phytoplankton or have come across images of zooplankton in a scientific exhibition or outreach publication. However, widespread awareness and appreciation of the vital role of plankton remain insufficient.

In this book, I will attempt to provide a scientific yet clear understanding of what plankton are and the various functions they perform. I will introduce you to creatures that may be unfamiliar, but could easily appear in fantastic or even terrifying novels, and discuss processes that occur in the sea every day, unnoticed by most. In short, I will delve into everything that is not typically seen when looking at the sea and what lies hidden within a mere droplet of water. Throughout, I will strive to present this information in an engaging and entertaining manner, while maintaining scientific rigor.

The book begins with some general chapters on the functioning of the marine planktonic food web, followed by details about specific groups, and curious aspects of plankton, and ends with a few chapters about plankton study and discovery. You can read this book from start to finish (which I recommend) or choose random chapters, although I suggest starting with at least Chap. 1.

Contents

About the Author

Albert Calbet is a marine researcher working at the Institute of Marine Sciences, CSIC (Barcelona, Spain), specialized in the ecology and eco-physiology of micro- and mesozooplankton. He has made significant contributions to the field, particularly in understanding the role of micro-zooplankton in marine food webs. Albert completed his Ph.D. in Marine Sciences in 1997 at the Institute of Marine Sciences (ICM), CSIC, after which he pursued postdoctoral research at the University of Hawaii at Manoa. He has held various positions at ICM, including Deputy Director. Albert has published over 120 peer-reviewed articles, authored books, and book chapters and actively participated in scientific conferences worldwide. He has been involved in teaching and mentoring students at the Ph.D., Master, and undergraduate levels. Albert's research has been supported by prestigious institutions, and he has served as a reviewer for funding agencies and on the editorial boards of scientific journals. Committed to science outreach, he manages several Web pages and engages with the public through social media, outreach articles, and books.

1

A Teaspoon of Seawater: A Tiny Ecosystem

© The Author(s), under exclusive license to Springer Nature Switzerland AG 2024
A. Calbet, *The Wonders of Marine Plankton*,
https://doi.org/10.1007/978-3-031-50766-3_1

The ocean's smallest inhabitants, plankton, drive vital processes. Phytoplankton, microscopic plant-like organisms, produce oxygen and fuel marine food webs. Bacteria break down organic matter, while viruses control microbial populations. Zooplankton, including copepods, support ecosystems by feeding on phytoplankton and transferring energy to higher trophic levels. This intricate web sustains life and nutrient cycling, showcasing the ocean's efficiency and balance.

During my college years, when studying ecology, I recall an interesting exam question: How many whales inhabit the Mediterranean? To solve the problem, we were provided data on primary production (phytoplankton output) and the average weight of a whale. Using a model of energy transfer through the trophic web, assuming 10% efficiency at each trophic level, the calculation itself was straightforward. The real challenge was determining the number of trophic steps to consider. At that time, when we were all naively confident, believing things unfolded exactly as described in textbooks, we settled for a trophic network model with two, maybe three, steps between algae and whales. However, after devoting more than 30 years to studying marine planktonic food webs, I now realize I would likely fail that exam. The problem lies in the fact that nature is far more intricate than we could ever anticipate, and generalizations are often arduous, if not impossible.

Typically, the smallest organisms are not only the most abundant but also play vital roles. This holds for plankton, which, though mostly invisible to the naked eye, are crucial to marine trophic food webs. Planktonic organisms sustain life on Earth, generating half of the oxygen produced on the planet and forming the basis of the nourishment of the fish we consume. Unfortunately, they also originated fossil fuels like oil. What can we do? After all, nobody is perfect.

Before digging into the story of plankton, let's introduce the primary characters. Imagine a teaspoon of seawater, approximately five milliliters, that you could collect at the beach. Within this small volume, we would find around 50 million viruses, five million bacteria, several hundred thousand photosynthetic or heterotrophic flagellates (unicellular organisms with whip-like appendages), thousands of microscopic algae, a handful of heterotrophic ciliates or dinoflagellates, and if we are lucky, one or two small crustaceans, such as copepods. The plant component of plankton is referred to as phytoplankton, whereas the animal component is known as zooplankton. Although the term zooplankton encompasses both unicellular and multicellular organisms, we typically distinguish them by size. Hence, we have microzooplankton (mostly unicellular and very small) and mesozooplankton (multicellular, or proper animals).

Each member of the plankton community serves a specific function. Viruses (which do not affect humans) regulate the populations of bacteria and other microorganisms to prevent excessive proliferation. Bacteria,[1] in turn, decompose organic matter and aid in nutrient recycling. Without them, corpses and waste would accumulate to unimaginable levels; considering the length of time we have inhabited this planet, if not for the activity of bacteria, we would likely be walking or even climbing upon a sea of corpses. Bacteria carry out intricate reactions that break down organic matter into inorganic compounds (salts of nitrogen, phosphorous, etc.), which can then be utilized by the primary producers—phytoplankton. Additionally, bacteria contribute to the cycling of various chemical elements, such as sulfur and iron.

Phytoplankton (marine unicellular vegetables; Fig. 1.1) are responsible for assimilating the inorganic salts generated by bacterial activity (particularly nitrates, phosphates, and silicates) and incorporating them into living matter. Through the process of photosynthesis, they also capture CO_2 (+ H_2O) and convert it into organic matter with the assistance of solar energy. During this photosynthetic process, phytoplankton release oxygen—a by-product that has the unfortunate tendency to oxidize objects but is essential for our existence. Concerning oxygen, it is frequently mentioned on outreach websites, in the press, and even in some scientific publications that marine algae contribute to roughly half of the oxygen we breathe. Although this statement may hold on a geological timescale spanning millions of years, it is not accurate on a daily basis. It is correct that phytoplankton have contributed to accumulating over half of the oxygen in the Earth's atmosphere throughout the history of the planet, and without their photosynthetic activity, animal life as we know

---

[1] Actually, the correct term should be prokaryotes, which encompass bacteria and archaea. However, for the sake of clarity I will use the term bacteria.

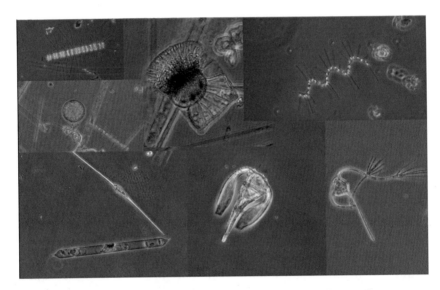

Fig. 1.1 Different microalgae from the phytoplankton. © Albert Calbet

it would not have colonized the Earth's surface. It is also true that phytoplankton, with only about 1% of the biomass of terrestrial plants, account for approximately 50% of photosynthesis and is responsible for roughly half of the daily oxygen production on the planet. However, the majority of this oxygen produced in seas and oceans is consumed by the various micro- and macro-organisms that reside there, and only a minute fraction reaches the atmosphere.

Consumers of phytoplankton can range from unicellular organisms like flagellates, ciliates (Fig. 1.2), dinoflagellates, or foraminifera, to multicellular ones, such as worm and mollusk larvae, starfish, fish larvae, crustaceans, and jellyfish. Among the multicellular organisms, copepods (Fig. 1.3), a group of crustaceans usually measuring no much more than a millimeter, are the most abundant animals on the planet—surpassing even insects. Copepods serve as the primary food source for fish (and occasionally whales, although they prefer larger prey like krill). However, before reaching this stage in the food web, copepods must feed on ciliates, algae, and other microscopic organisms.

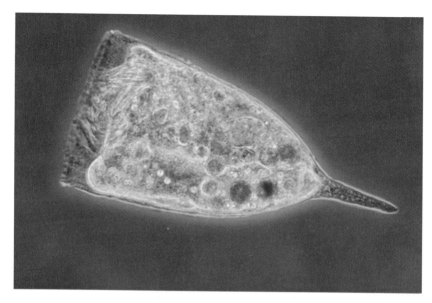

Fig. 1.2 Tintinnid ciliate. © Albert Calbet

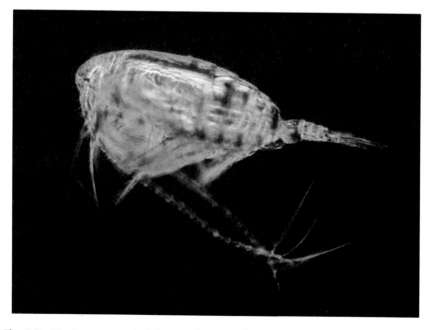

Fig. 1.3 Marine copepod. *Calanus minor.* © Albert Calbet

Likewise, ciliates also consume algae, flagellates, or bacteria, while flagellates can feed on bacteria or other flagellates. This interplay of interactions forms an intricate net known as a trophic food web (Fig. 1.4 above). As the members of the food web consume each other, they release the nutrients stored in their living matter, making them available to algae in an essential process known as nutrient recycling and conducted within the microbial loop (Fig. 1.4 below). It is fascinating to realize that we often perceive recycling as a modern concept, but in reality, it has been occurring in the sea for countless millions of years. In fact, very little goes to waste in the ocean, almost everything finds a purpose.

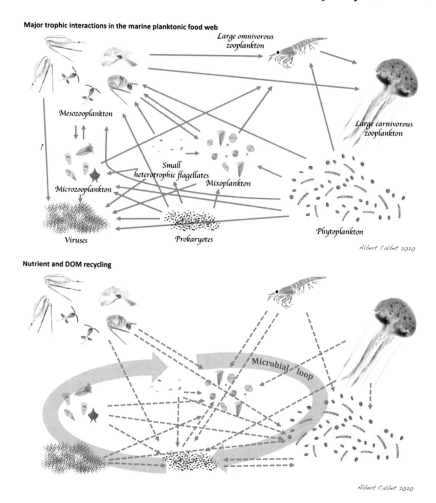

Fig. 1.4 Marine food web scheme. Above: trophic interactions, bellow: nutrient and dissolved organic matter recycling within the microbial loop. Note that bacteria are included within the group prokaryotes. © Albert Calbet

2

The Marine Biological Pump

The biological pump, orchestrated by minute phytoplankton, removes CO_2 from the atmosphere, depositing it in ocean sediments for centuries. This vital process mitigates global warming by capturing 25–50% of the CO_2 produced by humans. Tracing a carbon atom's journey, from diffusion into the ocean to sinking into the sediment, illustrates the intricate mechanisms sustaining marine life.

The biological pump is a process through which the ocean, with the assistance of marine organisms, removes carbon dioxide (CO_2) from the atmosphere and deposits it in sediments, where it will remain for hundreds or even thousands of years. This mechanism plays a crucial role in mitigating the effects of global warming, as it captures and incorporates 25–50% of the CO_2 produced by human activity. Surprisingly, this immense task is primarily carried out by minute unicellular organisms known as phytoplankton. Phytoplankton consist of microalgae, measuring only a few thousandths of a millimeter (micrometers), but their significance lies in their role as the driving force behind photosynthesis in marine planktonic food webs.

To grasp the functioning of the biological pump, imagine yourself as a carbon atom, accompanied by two oxygen atoms, forming a CO_2 molecule—the notorious by-product of burning fossil fuels (Fig. 2.1). Perhaps you originated from a car exhaust pipe, an industrial chimney, or even from the exhalation of your neighbor—it does not matter. In your carbon form, you joyfully traverse the boundary of the sea by a process called diffusion. In the water, you are swiftly trapped by a small microalga, which, aided by the energy of the sun and inorganic nutrients, transforms you into living matter. Although you bask in the pride of being part of a larger and more organized entity than a mere molecule (now forming a sugar chain), your joy is short-lived. A mixotrophic dinoflagellate engulfs you, disintegrating the complex structure you had become into smaller fractions, utilizing them to construct other intricate components. Yet again, your fate takes an unexpected turn as a passing ciliate incorporates you into its diet, restarting the digestion process. However, this time, fortune smiles upon you as a copepod chews the ciliate, and you find yourself within the digestive tract of this creature. Through time and the patient workings of catabolic and anabolic

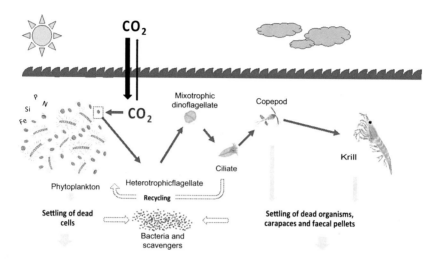

Fig. 2.1   Representation of one of the multiple paths of the biological pump. ©
Albert Calbet

processes, you become part of a lipid chain that resides in the cephalothorax of the copepod, encapsulated within a droplet of fat. As your host
migrates to deeper zones during the day to evade visual predators like
fish, you ascend to shallower layers at sunset, where algae and other prey
thrive. However, a krill attacks, causing the copepod to split in two, with
your portion escaping ingestion. Gradually, you settle into the depths of
the ocean, attached to a piece of sea snow, as bacteria and other microorganisms commence the decomposition of the remains of the copepod,
progressively reducing your holding structure into smaller fragments.
Suddenly, there is an abrupt change in speed as multiple decomposing
particles aggregate, causing you to descend more rapidly, affixed to a cluster of sea snow. After what seems like days, you finally reach the ocean
floor, where there remains a slight chance of reentering the benthic trophic network through the actions of crabs, worms, or other creatures.
However, whether due to chance or the recalcitrant characteristics of
your molecule, your carbon atom remains undisturbed, gradually sinking
deeper into the sediment, where you will remain for years, decades, centuries, or even millennia. In this manner, the carbon atom that once
belonged to a CO_2 molecule becomes trapped in the depths of the ocean.

The pathways by which a carbon atom embarks on its journey from the atmosphere to the ocean floor are limitless and exhibit a wide range of durations. Some pathways, like the one illustrated here, may last only a few days, while others may extend for hundreds or thousands of years, or perhaps never reach their final destination at all. Regardless of the specific trajectory, each carbon atom plays a vital role in the intricate mechanisms that sustain life in our oceans. From its initial existence as a CO_2 molecule, it undergoes a remarkable transformation, becoming incorporated into the food web of marine organisms, ultimately finding its resting place in the depths of the sea.

3

The Four Seasons of Plankton

A. Calbet, *The Wonders of Marine Plankton*,
https://doi.org/10.1007/978-3-031-50766-3_3

Plankton's harmonious dance with the seasons forms the backbone of marine life. From winter nutrient-rich mixed waters to summer stratified layers, the intricate interplay of biology, physics, and chemistry drives cycles of growth and survival. These cycles, disrupted by climate change, emphasize the ecosystem's fragility and importance. Just as Vivaldi's music captures the essence of seasons, so does plankton encapsulate the wonders of nature.

Vivaldi's Four Seasons, composed in 1721, are among the most famous classical music pieces of all times. By listening to each of the concerts, we are instinctively transported into the splendor and severity of nature. Similarly, plankton live and vibrate in harmony with the changing seasons, repeating year after year. In this chapter, I will provide an overview of the seasons in a standard temperate ecosystem. I hope that the next time you gaze upon the sea, whether it be summer, fall, winter, or spring, you can imagine how the little creatures that reside there interact, relate, and struggle for survival. You will discover that in the sea, biology, physics, and chemistry are intertwined, and one cannot be fully understood without considering the others. These few lines aim to foster an appreciation for the fragile and invisible plankton ecosystem, which sustains us by serving as the foundation of the marine food web.

To aid comprehension, I have prepared a small drawing summarizing the key aspects of each season (Fig. 3.1). We will begin with winter, when the water turns cold and is stirred by the wind, while the sunlight, barely rising above the horizon, weakly penetrates the surface of the ocean. Despite the abundance of nutrients resulting from the intense mixing of surface and deep waters, the limited availability of light and low temperatures restricts the growth of phytoplankton. As winter gives way to spring, the light becomes more intense, and the temperature begins to rise. The gradual warming of the water forms a thin thermocline, separating the surficial mixing layer from deeper areas. These conditions facilitate phytoplankton blooms, followed by an increase in microzooplankton populations and subsequently larger zooplankton (e.g., copepods). As time passes and temperatures continue to rise, summer arrives. During this season, a well-defined thermocline divides the ocean into two distinct areas: a shallow, nutrient-depleted region

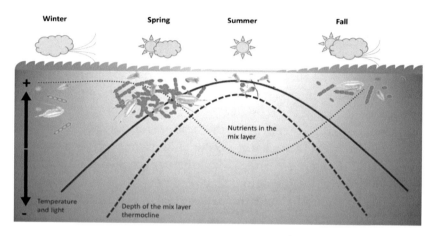

Fig. 3.1 Scheme of the four seasons of plankton in a temperate ecosystem. © Albert Calbet

(approximately 50 meters deep in open waters) due to the consumption of nutrients by phytoplankton in spring and early summer, and a deep, cold, nutrient-rich zone. In the early days of summer, nutrient-limited phytoplankton make way for a unique community of algae adapted to these conditions. These summer algae are either small in size, with a high surface-to-volume ratio that allows efficient use of the limited nutrients, or they are large and motile (e.g., dinoflagellates) capable of exploring micro-patches of remaining nutrients. Under favorable circumstances, these dinoflagellates can form harmful algal blooms, formerly known as "red tides" (although they are not always red, nor are they tidal in nature). Consumers within these summer communities include microzooplankton, small-particle filter-feeding zooplankton like gelatinous organisms or marine cladocerans (water fleas), as well as some carnivorous predators such as certain copepod species, and detritivores. When autumn arrives with its storms and intensified winds, the thermocline breaks down, allowing nutrient-rich waters to reach the surface. In some cases, there may be another small phytoplankton bloom if weather conditions are favorable, but often, low light and temperatures prevent the phytoplankton from fully utilizing the fresh nutrient input. And so, winter returns, and the cycle begins again.

As you can see, life cycles repeat year after year in the realm of plankton. Many marine organisms have developed internal clocks that guide their behaviors, such as when to lay eggs, so the newly hatched larvae would coincide with their prey blooming. Unfortunately, with the onset of climate change, these delicate rhythms are disrupted, impacting the natural functioning of the ecosystem.

In conclusion, let us appreciate the beauty and intricacy of the four seasons of plankton. They parallel Vivaldi's musical masterpiece, connecting us to the wonders of nature. May this understanding inspire us to cherish and protect the fragile and vital plankton ecosystem, recognizing its role in sustaining life and the delicate balance it maintains within our oceans.

4

The Rhythms of Plankton

The rhythms of nature shape the world, from the daily cycle of sunlight to larger patterns like seasons and climatic phenomena. Plankton, fundamental to marine ecosystems, align with these rhythms. Circadian rhythms drive phytoplankton photosynthesis and zooplankton migration, while circalunar rhythms link zooplankton depth to moonlight. Circannual rhythms synchronize spawning, migrations, and the plankton bloom. Multi-year cycles like El Niño impact nutrient-rich upwelling and food webs. These rhythms orchestrate nature's symphony, guiding the dance of life in the oceans.

The dichotomy between day and night serves as the driving force behind many rhythms in nature. The presence of the sun triggers thousands of biochemical, physiological, and ethological processes through photosynthesis. While humans predominantly sleep at night, there are numerous creatures that thrive in darkness. During the night, we observe the moon, which also operates on its own rhythm of approximately 28–29 days. The duration of daylight and darkness in different latitudes is influenced by the seasons, which repeat with varying degrees of accuracy each year. In essence, our lives are shaped by rhythms. From the daily news broadcasts to Yogi Bear's winter hibernation, or even the appearance of werewolves and other lunatics. All these occurrences exhibit repetitive patterns. Plankton, as one would expect, are similarly deeply influenced by rhythms of varying amplitudes and intensities. While I will provide a few examples, it is important to note that I do not aim to be exhaustive, nor do I wish to delve into extensive details about the triggering mechanisms, simply because many of them remain unknown. In most cases, the rhythm is regulated either by external factors that adjust on a daily basis (such as hours of daylight) or, due to evolutionary processes, an internal clock has been established that functions independently of the presence or absence of the light.

Circadian Rhythms

Phytoplankton, unicellular planktonic algae, actively undergo photosynthesis during the day and respire at night. Consequently, many species take advantage of the darkness to divide and reproduce. During the night, large zooplankton organisms like copepods and krill migrate from the

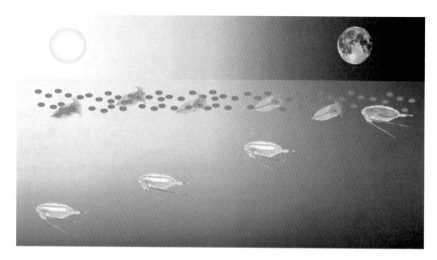

Fig. 4.1 Drawing showing vertically the daily migration of zooplankton. ©
Albert Calbet

dark, deep regions of the ocean to the surface to feed on phytoplankton
(Fig. 4.1). By feeding at night, zooplankton members avoid being seen
and attacked by their predators, such as fish. Copepods also consume
microzooplankton, which occupy the same depths as the algae.
Microzooplankton primarily feed on algae during the day to minimize
predation by copepods, which are absent during that time. As you can
see, everything operates in an orderly and balanced manner, largely due
to the millions of years of coevolution between predators and prey.

Circalunar Rhythms

In the Arctic night and presumably in the Antarctic Ocean as well, the
depth at which zooplankton are found is influenced by the illumination
of moonlight. The brighter the moon, the deeper the zooplankton reside.
This behavior helps them avoid predation by fish adapted to low-light
conditions. Similar moonlight effects occur in other oceans, modulating
the depth at which zooplankton are found at night. Even lunar or solar
eclipses disrupt the migratory patterns of zooplankton.

Circannual Rhythms

The four seasons serve as another example of periodicity that is more or less accurately repeated each year. As expected, the different weather conditions associated with the length of day and night in each season shape the dynamics of the marine ecosystem. Synchronous spawning of corals or polychaetes, as well as whale migrations, are some examples among the countless instances found at sea. However, perhaps the most significant example, given its global relevance, is the succession of plankton organisms throughout the seasons. This succession, combined with the physicochemical characteristics of the water associated with each season, is responsible for the spring (at times late winter) phytoplankton bloom. This bloom, in turn, supports a thriving zooplankton community that serves as a food source for fish and other marine creatures.

Multi-Year Cycles

Although not strictly referred to as rhythms, there are significant climatic phenomena that repeat every few years. One of the most well-known is "El Niño," which primarily affects the Pacific Ocean but also has global consequences (Fig. 4.2). "El Niño" and its counterpart, "La Niña," are cyclical variations in temperature that typically occur every 4 years (alternating between up and down) in the central and eastern tropical regions

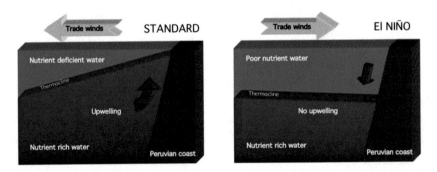

Fig. 4.2 Representation of "El Niño" phenomenon. © Albert Calbet

of the Pacific Ocean. These phenomena have implications beyond changes in temperature and rainfall; they directly impact the upwelling of nutrient-rich deep waters along the west coast of the American continent. Under "El Niño" conditions, the trade winds that normally facilitate upwelling reverse their direction and weaken, leading to consequences for phytoplankton that rely on these nutrients and for the entire food web that depends on them, including fish. These effects have serious socio-economic implications for the affected areas.

Another similar phenomenon is the North Atlantic Oscillation (NOA), characterized by pressure changes between the Azores and the subpolar zone of the North Atlantic. Positive oscillations result in higher temperatures in northern Europe and usually the opposite in southern Europe. The Gulf Stream is also affected, along with a range of planktonic species and fish that rely on it.

The ocean harbors numerous rhythms and cyclical processes, each with its unique characteristics and idiosyncrasies. Often, we may not be aware of these rhythms, but nature always moves in harmony with its own cadence.

5

Plankton, Mussels with French Fries, and Diarrhea

Plankton, usually harmless, can include toxic phytoplankton species that cause harmful algal blooms (HABs). These blooms, often referred to as "red tides," result from factors like physical accumulation and excessive growth, posing threats to marine life and humans. Toxin-producing species affect various organisms, leading to mass deaths, altered behavior, and even human health issues. Human activities such as global warming and nutrient pollution exacerbate these phenomena. Monitoring efforts by experts help mitigate the risks posed by these toxic algae.

The Belgians have the culinary tradition of eating mussels with French fries—it is all a matter of taste. In Spain, on the other hand, they serve them alone, in sauce, or as a complement to paellas or other equally delicious dishes. Mussels, apart from being a gastronomic delight available to everyone, are incredibly efficient filtering machines. In their daily activity, they can filter and clear up to 200 liters of seawater per day of plankton. And this is where the problems begin. Plankton are composed of a multitude of organisms, which do not normally pose any danger to humans. However, certain groups of phytoplankton are toxic. These include many dinoflagellated algae (Fig. 5.1) and a few species of diatoms and cyanobacteria. At the concentrations we usually find them, nothing happens if we accidentally swallow a sip of seawater. However, if they are concentrated by a filter-feeder, such as mussels or oysters, or if they have climbed the food web and bio-accumulated in fish, they can be harmful to humans and even cause death.

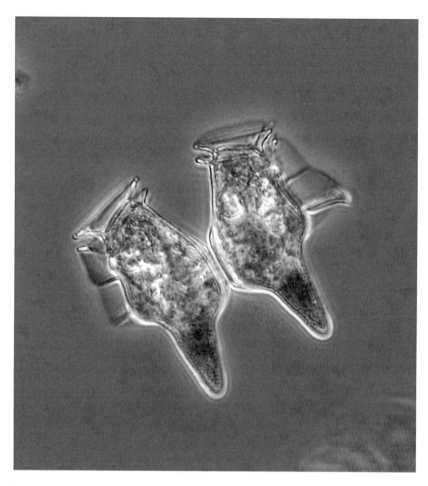

Fig. 5.1 *Dinophysis* sp. one of the most toxic dinoflagellates. © Albert Calbet

What Are the So-Called "Red Tides"?

They are nothing more than accumulations of unicellular algae, usually toxic, in a certain area to a greater or lesser extent. Many of these algal blooms indeed have a reddish coloration, but there are several other colors, such as brown or green, which makes it strange to call them red tides when the water is, for example, green. Therefore, to be more precise, scientists have long adopted the term "harmful algal blooms" (HABs). These proliferations differ from the typical spring blooms because they usually adversely affect the ecosystem as a whole or at least one or another organism, including us.

Why Do HABs Occur?

Well, unfortunately, the origin of HABs is not clear. They may occur due to physical accumulation in semi-confined areas, such as ports or estuaries, promoted by sea currents, or due to exaggerated growth of some species that can suddenly take advantage of the opportunity of ideal conditions, or by the release of new cells from cysts accumulated in the sediment. The fact is that, more or less, every summer (and even some winters) we have some of these proliferations on our shores. What is certain is that human activity contributes to the increase of these phenomena. Global warming is accelerating their growth rates, and also, marinas act as centers of aggregation and initial seed for proliferations. Moreover, the release of excessive nutrients into the sea from city sewage, agricultural activity, etc., provides the necessary food for these algae and also promotes HABs. All in all, we are helping to create growing broths suitable for HABs. Fortunately, some of these issues (such as sewer efficiency and control over agricultural fertilizer) are currently being addressed, and the results are promising.

How Do They Affect Marine and Non-marine Life?

There are many types of toxic algae and toxins. Some directly attack mollusks and fish (among other organisms), causing massive deaths in nature and aquaculture facilities. Others, such as the diatom *Pseudonitzschia*, which produces a potent neurotoxin (domoic acid), can drive animals crazy. Remember Alfred Hitchcock's "The Birds"? It is believed that the famous director was based on a true case of affected seagull after eating fish poisoned by this diatom. There are also cases of sea lions influenced by this toxin that act like zombies. Other HABs enter the food web from its base, acting on zooplankton and other algae species. Some are harmless, and their effects are simply aesthetic and harmful only to tourists looking for crystal-clear water on the beaches. The worst, however, from a human point of view, are those that go more or less unnoticed through the food web and, ultimately, reach us. The range of effects of these organisms on our species includes well-known summer diarrhea, vomiting, convulsions, and even death. A rather curious and unfortunate case happened in a laboratory in North Carolina in 1993, where two scientists who cultivated *Pfiesteria*, a toxic dinoflagellate that causes fish mortality, progressively showed symptoms of memory loss and disorientation. They never completed their study, and to date, they still have sequelae in their cognitive abilities.

As you can see, the risks from the proliferation of toxic algae are considerable. Fortunately, regular monitoring programs have been established in places that may be affected by these algal blooms, especially near aquaculture facilities. Currently, monitoring consists of a handful of experts who examine water samples under a microscope and identify the usual suspects. Surely, soon this will be done automatically with machines that perform real-time molecular or chemical analysis. But for now, we have to trust our experts.

6

Are There Carnivorous Plants in the Sea?

A. Calbet, *The Wonders of Marine Plankton*,
https://doi.org/10.1007/978-3-031-50766-3_6

Carnivorous plants have evolved in unique environments, extracting nutrients from insects and creatures through traps. In the marine realm, mixotrophic microalgae perform similar roles, consuming prey like other algae. Some even acquire plant-like characteristics, incorporating chloroplasts or entire algae for photosynthesis. Mixotrophy's role in marine trophic food webs remains a topic of study, with each species holding unique behaviors and significance.

Whether it is Audrey from Frank Oz's musical "Little Shop of Horrors," John Wyndham's "The Day of the Triffids," or the small Venus flytrap plants sold in flower shops, everyone knows what carnivorous or mixotrophic plants are. You might have even tried to grow one at home, probably with little success. This is because they are plants adapted to very specific environments, typically characterized by acidic soils with low nutrient content, high humidity, and precise temperature ranges. However, with these conditions in mind, keeping a few species at home is not too difficult.

Due to the unique environments in which these plants thrive, they have evolved mechanisms to extract nutrients that are lacking in the soil from insects and other creatures. They attract and capture these organisms using modified leaves in the form of various traps. However, if the environment deviates even slightly from their specific requirements, these plants either die or are quickly outcompeted by faster-growing species. As a result, carnivorous plants are found in only a few locations on Earth, while non-carnivorous, strictly photosynthetic plants are ubiquitous.

But what about the sea? Well, there are numerous mixotrophic plants in the sea that consume other organisms. The reason they are not as well-known is because they are unicellular and invisible without a microscope. Apart from diatoms (algae with siliceous skeletons) and a few representatives from other groups, most planktonic microalgae can feed on live prey. Imagine if almost every plant on Earth were carnivorous—there would be no insects left!

In the marine environment, the range of prey available to these mixotrophs is quite broad. Most of them consume other algae to obtain inorganic nutrients like nitrogen or phosphorus, replenish their chloroplasts, as a carbon source, or to eliminate competition for resources (Fig. 6.1).

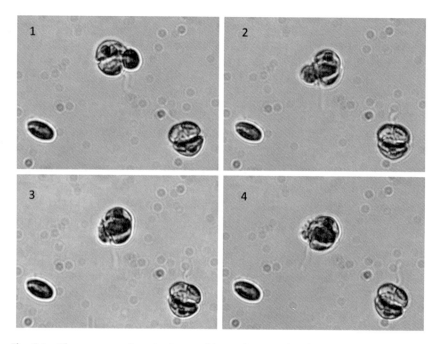

Fig. 6.1 The process of capturing and ingesting an alga (*Rhodomonas salina*) by a mixotrophic dinoflagellate (*Karlodinium veneficum*). © Albert Calbet

Many marine mixotrophs, although primarily vegetarian, do not hesitate to capture animal prey, be it unicellular or multicellular. They immobilize and kill their prey using venomous spears or by releasing toxins into the water.

In addition to these mixotrophs, we also find a unique type of organism in the marine ecosystem that does not exist on land, except in science fiction films and comics like "The thing from another world" or "The swamp thing." These are "animals" with plant-like characteristics, known as non-constitutive mixotrophs. They acquire these characteristics by capturing plant prey and incorporating their chloroplasts (or the entire algae) to perform photosynthesis. Can you imagine a rabbit as green as lettuce? In reality, a few marine multicellular animals possess this ability as well. Some corals, sponges, worms, and bivalves capture symbiotic algae. Even the green sea slug *Elysia chlorotica* can synthesize rudimentary chlorophyll. However, unicellular organisms are the true kings of non-constitutive mixotrophy in the sea. Ciliates, dinoflagellates, foraminifera, radiolarians, and other groups are capable of capturing and assimilating whole algae or their chloroplasts. Some even incorporate the nucleus of the captured cell into their cytoplasm to aid in the duplication process—a remarkably complex feat for a single cell. Certain species exhibit high levels of specialization, preying exclusively on one specific prey species or even one particular mixotroph species that has previously acquired chloroplasts from a specific prey group. For example, the dinoflagellates of the genus *Dinophysis* feed on the ciliate *Mesodinium rubrum*, which, in turn, consumes and captures chloroplasts from a particular group of algae. The plasticity and ease of incorporating foreign organelles from many marine protists is likely a result of the evolution of eukaryotic cells in the sea. It is believed that the origin of algae began with a cell lacking photosynthetic capacity that was able to capture and retain an autotrophic bacterium, the first chloroplast. This event occurred approximately 1500–2000 million years ago, at the dawn of life on Earth.

As you can see, despite being a widespread phenomenon, we still have much to learn about the factors involved in mixotrophy in the sea. Each species is a unique world, and even different strains of the same species may exhibit distinct behaviors. We are still uncertain about the true significance of mixotrophs in marine trophic food webs because possessing the ability to use a particular metabolic pathway does not necessarily imply its utilization.

7

The Most Formidable Predators of the Ocean: The Protists

A. Calbet, *The Wonders of Marine Plankton*,
https://doi.org/10.1007/978-3-031-50766-3_7

Marine protists, single-celled eukaryotic organisms, are formidable predators with diverse feeding strategies. Protozoa and mixotrophs consume a significant portion of carbon produced by algae in the ocean, employing methods such as filtration, engulfment, tube feeding, pallium feeding, and even piston-like mechanisms. Despite their tiny size, these creatures can immobilize, ingest, and digest prey ranging from bacteria to copepods. Their adaptations and abilities highlight the incredible diversity of marine life.

"Mr. Vaughn, what we are dealing with here is a perfect engine, an eating machine. It's really a miracle of evolution. All this machine does is swim and eat and make little sharks, and that's all." This is how oceanographer Matt Hooper (Richard Dreyfuss) described the shark in Steven Spielberg's 1975 film Jaws.

When we think of efficient and fearsome marine predators, sharks often come to our mind, particularly the great white shark. However, had Mr. Spielberg known about the terrifying ways marine protists hunt and devour their prey, perhaps the film would have been titled "The Protist." Jokes aside, if there are any creepy and dangerous creatures lurking in the ocean, they are the protists.

Protists are ubiquitous single-celled eukaryotic organisms (i.e., with a nucleus) found in seas and oceans. They can be classified in various ways, such as by their energy acquisition methods: those that undergo photosynthesis (autotrophs or algae), those that consume other organisms (heterotrophs, also known as protozoa), and those that combine both strategies (mixotrophs). In this discussion, we will focus solely on the predators that feed on live prey, namely protozoa and mixotrophs.

Out of the approximately 50 gigatons (50 billion tons) of carbon produced annually by algae in the seas and oceans, protozoa (assisted by mixotrophs) consume about 30–60%. In comparison, copepods, the next significant consumers, consume only around six gigatons. Protozoa and mixotrophs not only prey on algae but also feed on bacteria, other protozoa, and even animals much larger than themselves. But how do these minute, mouthless unicellular organisms devour their prey? The truth is that they employ various prey capture and feeding strategies, each with its own peculiarities.

Major Feeding Strategies in Protists

Filtration Many microorganisms utilize feeding mechanisms similar to those of whales, either by attracting prey into their oral openings using feeding currents or by actively swimming and collecting them. This strategy is typically employed for capturing very small prey such as bacteria or flagellates. Engulfment As prey size increases, numerous protists can capture and ingest them whole, akin to a boa swallowing a goat. For example, the dinoflagellate *Gyrodinium dominans* possesses a highly flexible cell body and can ingest chains of diatoms much larger than itself (Fig. 7.1). Some foraminifera, distant relatives of amoebas protected by an outer covering made of calcium carbonate, can even consume large copepods (Fig. 7.2). Although this process is slow, it proves effective. Many protists employ venomous stingrays or release toxins into the water to immobilize their prey. Some toxins are so potent that they can kill fish and other organisms. Moreover, when these toxins accumulate in filter-feeding creatures like mussels, they can cause severe poisoning in humans. Tube or Peduncle Feeding Certain dinoflagellates possess a retractable tubular structure that they insert into their prey, sucking out its contents like a straw in a margarita cocktail. They

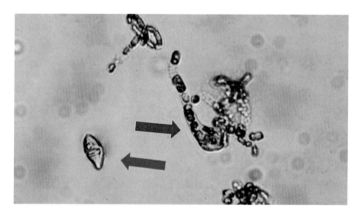

Fig. 7.1 Process of swallowing a chain of diatoms by the dinoflagellate *Gyrodinium dominans*. The red arrow indicates a *G. dominans* with a chain of diatoms inside. The blue arrow shows the size of the same species without prey inside. © Albert Calbet

Fig. 7.2 Foraminifer that just captured two copepods. © Albert Calbet

use this mechanism to consume prey similar in size to themselves, as well as to immobilize and devour animals much larger than them, such as copepods and worms. Pallium Feeding This feeding strategy is perhaps the most intriguing and complex. Like sea urchins and starfish, protists extend their stomachs (or rather, a membrane with digestive characteristics) to gradually digest their prey and incorporate the dissolved nutrients into their membrane, resembling a slow digestive process. For instance, they can consume large chains of diatoms using this method (Fig. 7.3). Once the trapped cells are consumed, only a siliceous skeleton remains. Piston In recent years, a peculiar dinoflagellate called *Erythropsidinium* was discovered. It possesses a small expandable piston that it can rapidly extend and retract. This piston is believed to be used for detecting and catching prey, which is eventually swallowed. What is even more fascinating about this single-celled organism is that it has a rudimentary eye called an ocelloid, complete with a lens. While the

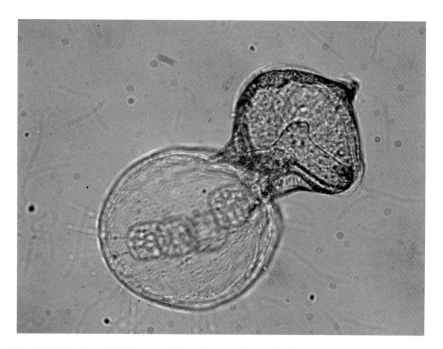

Fig. 7.3 *Protoperidinium* sp. dragging a chain of diatoms into its pallium. ©
Albert Calbet

exact function of this ocelloid is still debated among scientists, it is
thought to aid in locating prey. It is truly remarkable considering we are
talking about a single-celled creature! Fortunately for us, all these creatures, which could easily be mistaken for nightmares straight out of a
Stephen King novel, are no larger than a few tens of thousandths of a
millimeter. Just imagine the chaos if they were our size!

In conclusion, while sharks may have gained notoriety as the most
dangerous predators of the ocean, it is important to recognize the incredible diversity and abilities of marine protists. Their various feeding strategies showcase their remarkable adaptations for hunting and consuming
prey. The world of protists is full of wonders and mysteries, underscoring
the vast complexity of life in the ocean.

8

Copepods: Good Things Come in Small Packages

© The Author(s), under exclusive license to Springer Nature Switzerland AG 2024
A. Calbet, *The Wonders of Marine Plankton*,
https://doi.org/10.1007/978-3-031-50766-3_8

Copepods are tiny crustaceans vital to marine ecosystems. They employ diverse feeding strategies, and their remarkable speed and daily migrations are impressive. Their reproductive methods involve pheromones and swimming patterns to find mates. After mating, fertilized eggs hatch into nauplius larvae, which mature into adult copepods. Copepods are essential in marine food webs, linking primary producers to higher trophic levels and fisheries. Notably, Sapphirina copepods display stunning color-changing iridescence to attract mates, showcasing nature's beauty and complexity.

Imagine a group of organisms more abundant than insects, so abundant that if we lined them up, touching each other by their antennae, they would reach the sun and back. Imagine that they are the primary source of food for fish, to the extent that without them, the ocean would likely be devoid of fish. Imagine also that they are the fastest animals on Earth and that they undertake the proportionally largest daily animal migrations on the planet. It may sound like science fiction, but these beings do exist—they are the copepods.

What Are Copepods?

Copepods, derived from the Greek words "cope" (paddle) and "poda" (leg), are a subclass of aquatic crustaceans that inhabit virtually all seas and oceans. They are small, ranging from less than a millimeter to almost a centimeter in length, with most species averaging around a millimeter. Their abundance is staggering, making them the most populous group of metazoans on the planet. With approximately 12,000 described species, copepods possess unique characteristics that make them essential for the functioning of marine food webs. While most copepods are free-living, there are also many parasitic species (Fig. 8.1) that infest other organisms such as mollusks, fish, and even cetaceans. These parasitic copepods exhibit bizarre shapes adapted to their specialized functions, differing greatly from their free-living counterparts, which resemble small shrimp with elongated abdomens and large antennae (Fig. 8.2). However, some copepods are flattened or elongated, appearing thin.

Fig. 8.1 Monstrillidae copepod. These are parasites in their younger stages. © Albert Calbet

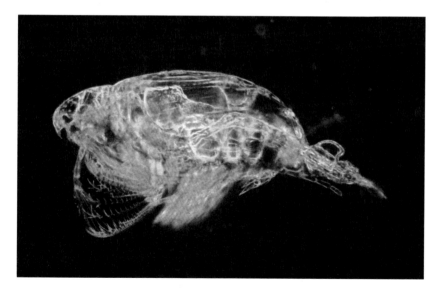

Fig. 8.2 Female of *Labidocera* sp. © Albert Calbet

The Fastest Animals

Copepods employ a combination of their antennae, five pairs of flat legs on the ventral part of their cephalothorax, and buccal appendages to propel themselves. They can achieve remarkable speeds, reaching up to 3–6 km h. Although this might not seem impressive at first, given their tiny size, it is equivalent to covering a distance a thousand times their body length in just 1 s. To put it into perspective, it would be like jumping over ten building blocks in a second! These high speeds are mainly attained during hunting or when evading predators, which they detect using the mechanosensors in their antennae. Copepods not only possess impressive short-term speed but also exhibit persistence in their displacements. Most species engage in daily migrations, traveling between 100 and 1000 m in just a few hours from the deep ocean, where they seek refuge from predators during the day, to the surface at night, where they feed on phytoplankton and microzooplankton.

Dietary Preferences

Due to their size, copepods exist between the realms physicists refer to as the viscous and inertial worlds. This means that although they perceive water as a liquid and fluid medium similar to how humans do, the prey they capture are so minuscule that, according to the laws of physics, their environment becomes much more viscous, akin to honey. As a result, movement, as we perceive it, becomes challenging, necessitating particular adaptations for both locomotion (flagella and cilia) and feeding. Copepods possess specialized structures near their mouths called mandibles, maxillae, and maxillipeds, which assist in capturing and consuming prey. The specific mechanisms employed depend on the copepod species and the size and motility of the prey. When dealing with larger prey, such as large protists or other copepods, they employ an ambush strategy, remaining motionless until an unsuspecting prey comes within reach, at which point they pounce and capture it. For smaller prey, copepods create currents while moving, attracting particles into their oral cavity, where the appendages mentioned earlier pick them up and, if suitable, transport them to the mouth. Until recently, the feeding process was believed to be simpler, with copepods filtering the water and indiscriminately consuming anything that entered their mouths. However, research conducted around the 1980s revealed that the mechanism is more complex, as I explained earlier.

Finding the Right Mate

Copepods have two distinct sexes (Fig. 8.3) and require copulation to reproduce. The reproductive process involves the production of eggs, which are either released into the sea or carried on the abdomen until the larvae, known as nauplii, hatch by breaking out of the eggshell. At first glance, this may seem straightforward, but the real challenge lies in finding a suitable partner of the opposite sex and the same species. Considering that the first 1000 m of the world's oceans (where copepods can be found) contain approximately 361 million km^3 of water, and that the average copepod is barely longer than a millimeter and lacks eyes, one can appreciate the difficulty involved in finding the perfect mate. It is quite a

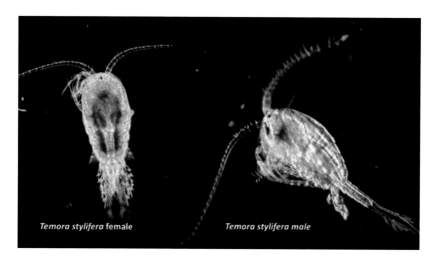

Temora stylifera female   *Temora stylifera* male

Fig. 8.3 Female (left) and male (right) of *Temora stylifera*. © Albert Calbet

daunting task! However, evolution has endowed copepods with strategies to aid them in this endeavor. Depending on the species, two primary strategies are employed to locate a mate: pheromones or swimming patterns.

Pheromones Pheromones are chemical substances that serve various functions in living organisms, including reproduction. Copepod males are no exception, and many species rely on detecting the pheromones released by females. This, however, is not an easy task since pheromones have a short lifespan, and even the slightest turbulence can dilute and dissipate the signal. Copepods possess special receptors in their antennae that enable them to locate and identify the pheromone molecules emitted by females of their own species. Once a chemical signal is positively identified, they follow the path of the female until they locate her. Swimming Patterns Some copepod species employ a more intriguing method to identify suitable females. Males swim continuously until they hydromechanically detect another copepod (also through receptors in their antennae). They then engage in a ritualistic dance characterized by small oscillations of a specific length and frequency. If the other copepod follows this rhythm, it indicates that she is the right female for copulation. If not, the male must restart the search process. This method of locating females is less efficient than pheromones and is typically observed in species with high abundances in the marine environment and relatively equal proportions of males and females. What happens when a copepod finds its perfect match? Once a compatible pair has been identified, copulation can take place. However, copulation is not a simple process for crustaceans, including copepods. In the case of copepods, males capture and restrain females using a modified antenna, typically more muscular than the others. With the female immobilized, the males employ the fifth pair of legs on their thorax, which is also highly modified and differs across species, to deposit a packet of sperm into the genital orifice of the female. It is truly a remarkable act of precision and coordination! Having described this intricate process, we can only hope for a positive outcome, with fertilized eggs being laid a few days later.

A Life From Larva to Larva

Once fertilized, females reproduce by laying eggs, which are either released into the water or carried until they hatch within a pouch located at the base of the abdomen, depending on the species. The duration of this process varies based on temperature and species. For example, at 20 °C, an egg of *Centropages typicus*—a common widespread temperate species—may take a day or two to hatch. From the egg emerges a larva called a nauplius (Fig. 8.4), which undergoes 11 molting stages before reaching adulthood—the twelfth larval stage. Once a copepod reaches adulthood, it ceases to grow and channels all its energy into feeding, evading predators, and reproducing. To provide some perspective, under the aforementioned conditions, the entire development process takes approximately 12–14 days, and an adult copepod can live for around a month or two. However, certain species adapted to polar environments require 2 years to complete their life cycle.

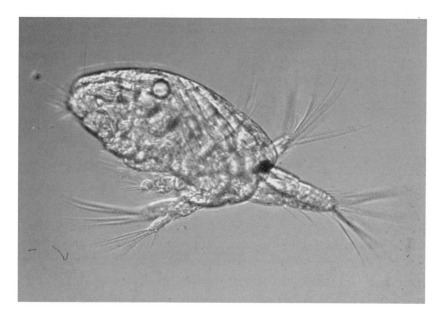

Fig. 8.4   Copepod nauplius. *Acartia grani*. © Albert Calbet

Why Are Copepods So Important?

The significance of copepods stems not only from the peculiarities of their anatomy and physiology but also from the crucial role they play in marine trophic food webs. As they feed on protists—both algae and protozoa—and serve as the primary prey for many fish species, especially during their larval stages, copepods act as a vital link connecting primary producers to fisheries. Estimating their abundance and production is essential for understanding and predicting future fishing stocks and regulating fishing efforts. Certain fisheries, like the cod fishery in the North Sea, rely exclusively on a single copepod species, *Calanus finmarchicus*. However, due to global warming, this cold-water species is shifting further north, and its former distribution areas are being occupied by other species with lower reproduction rates and potentially different nutritional qualities. This phenomenon is likely to have implications for the future of cod fishing. Copepods are also crucial as fish food in aquaculture and recreational fish farming. Their lipid composition, which includes high levels of omega-3 and other essential nutrients, is unmatched in nutritional quality. It has been demonstrated that larvae of delicate fish species have significantly higher survival rates when fed a diet primarily composed of copepods. Unfortunately, cultivating copepods in large numbers is challenging. Consequently, numerous laboratories are currently researching new cultivation methods for this fascinating group of organisms.

Sapphirina, an Astonishing Show of Light and Color

The *Sapphirina* (Fig. 8.5), a copepod that has earned the title of the world's most beautiful animal, deserves special mention in this chapter devoted to copepods. These hypnotic creatures are parasitic, targeting gelatinous plankton, particularly doliolids, as their hosts. While this predatory behavior alone is awe-inspiring and worthy of discussion, *Sapphirina* possesses additional peculiarities that set it apart not only

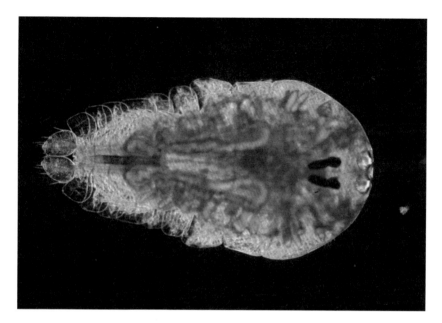

Fig. 8.5 *Sapphirina* sp. © Albert Calbet

within the realm of plankton but also within the animal kingdom as a whole.

To begin, the segmented body of the *Sapphirina* is incredibly flat and paper-thin, translucent in appearance. However, in a breathtaking display, it transitions from complete transparency to exhibiting iridescent hues spanning the entire spectrum of the rainbow. In fact, the name *Sapphirina* is derived from "sapphire," a fitting moniker given their resemblance to these precious gemstones. These captivating color changes are facilitated by hexagonal guanine crystals, a chemical compound employed by most cells in the production of DNA and RNA. While other organisms, such as chameleons, also utilize guanine crystals to alter their coloration, the ability of *Sapphirina* to rapidly and intensely shift hues within seconds is truly remarkable.

The purpose of this iridescence is still a subject of debate among the scientific community. The prevailing hypothesis suggests that since only males possess the capability to change color and females possess more developed eyes, this dazzling display serves as a mechanism to attract potential mates. In the case of *Sapphirina*, it appears that females are drawn to the mesmerizing light and color exhibition presented by the males. After all, who could resist being captivated by such a magnificent spectacle?

The *Sapphirina* copepod exemplifies the astonishing convergence of biological adaptations and aesthetic beauty. Their ability to transform into living kaleidoscopes provides a fascinating insight into the wonders of nature and the intricate complexities that underlie the natural world. The next time you encounter the extraordinary *Sapphirina*, take a moment to appreciate the breathtaking show of light and color that unfolds before your eyes.

9

Water Fleas, Matriarchy, and a Parasite Named Like a German Sausage

A. Calbet, *The Wonders of Marine Plankton*,
https://doi.org/10.1007/978-3-031-50766-3_9

Cladocerans, also known as water fleas, are abundant planktonic crustaceans in freshwater ecosystems. They help control algae and serve as food for other organisms. In the sea, they are found in warm and stable zones and reproduce parthenogenetically. Penilia avirostris is a common marine filter feeder, and genera like Podon and Evadne are predators. Their reproduction involves both clones and males, with resting eggs for survival. Parthenogenesis, a form of reproduction without males, is not unique to cladocerans and is observed in various species due to factors like infections with parasitic bacteria like Wolbachia. This bacterium has practical applications in pest control.

We can confidently say that in puddles, lakes, and reservoirs, the most abundant planktonic crustaceans are cladocerans, also known as water fleas. These iconic members of the zooplankton (which, beyond their shape, have nothing to do with terrestrial fleas) are known to some people because many kids (including myself at a young age) kept some at home to observe how they swim and reproduce.

The most abundant genus of freshwater cladocerans is *Daphnia* (Fig. 9.1 right), which comprises many species that are similar to each other. There are other less-known genera, though often abundant, such as *Bosmina, Moina, Cercopagis, Chydorus*, etc. Their roles in the ecosystem are primarily to control the proliferation of algae on which they feed and serve as food for other organisms, such as fish and insect larvae. Cladocerans are so numerous in freshwater that they have displaced copepods, which are mostly relegated to littoral and benthic zones in these ecosystems and have little relevance in the water column.

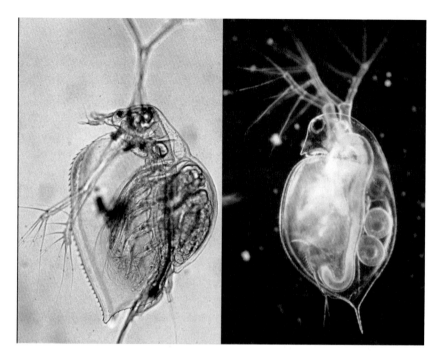

Fig. 9.1 Marine (left) and freshwater (right) cladocerans. © Albert Calbet

What Happens in the Sea?

In the sea, there are also cladocerans, but with less diversity and restricted to warm and stable zones of the water column. Thus, in tropical seas and temperate ones at summertime, one can find *Penilia avirostris* (Fig. 9.1 left); the only known species of the genus) as a filter feeder, and the genera *Podon* and *Evadne* as predators. It is believed that cladocerans are very sensitive to turbulence, which is why they only appear in periods of water column stability in the sea. It is also a hypothesis that continuous predation by fish and other marine organisms makes it impossible for stable populations of cladocerans to settle year-round, unlike in freshwater where they can proliferate if there are not many fish present.

An Army of Female Clones

The reproduction of cladocerans is quite curious. During periods of expansion, when environmental conditions are favorable, they reproduce parthenogenetically, meaning that females produce clones of themselves without any males involved. When things are going wrong, such as before drought in lakes or puddles, or toward the end of summer in the sea, males start to appear in the cladoceran population, fertilizing the females, who will mainly produce resting eggs called ephippia. These ephippia will settle and become buried in the sediment, remaining dormant until the next season or until conditions become favorable.

Parthenogenesis is not exclusive to cladocerans. Worms, tardigrades, crustaceans, insects, amphibians, reptiles, and even some fish and birds are examples of groups where species reproduce through parthenogenesis. The origin of this peculiar reproduction in each group is still a topic of debate. Perhaps the most interesting case is that of many arthropods (excluding cladocerans precisely) and nematodes, where the origin of parthenogenetic reproduction is due to an infection with a parasitic bacterium called *Wolbachia*. When this bacterium infects a host, it can have various actions: killing males, increasing the number of infected females in the offspring, creating parthenogenetic clones of females, or causing cytoplasmic incompatibility (infected males can only reproduce and have offspring with infected females). *Wolbachia* also has practical uses as a biological weapon to control pest organisms, such as mosquitoes that carry diseases. It has been observed that mosquitoes infected with *Wolbachia* have a reduced ability to spread viruses like dengue, Zika, or yellow fever.

In summary, this bacterium is a marvel of nature that we hope never mutates to infect humans.

10

Gelatinous Plankton: Itchy Jellyfish, Appendicularians with Luxury Chalets, Filter Barrels, and Other Extraordinary Creatures

A. Calbet, *The Wonders of Marine Plankton*,
https://doi.org/10.1007/978-3-031-50766-3_10

Jellyfish are part of gelatinous plankton, along with other organisms like chordates, fish eggs, and seaweed. This chapter focuses on metazoan zooplankton with jelly-like body structures, including appendicularians, pyrosomes, salps, doliolids, jellyfish, and ctenophores. The chapter ends with Phronima, a crustacean predator related to jellyfish, which carves out host organisms to use them as nurseries. Interestingly, it seems that Phronima inspired the terrifying creature in the movie "Alien."

Jellyfish are perhaps the most popular group of plankton. They cluster together with other organisms to form what we call gelatinous plankton. In this mixture, we can find chordates such as appendicularians or salps, fish eggs, jellyfish, and even some seaweed. However, in this book, we will focus on the metazoan zooplankton groups whose body structure is composed of jelly-like substances. The main groups are appendicularians, pyrosomes, salps, doliolids, jellyfish, and ctenophores.

Appendicularians

Appendicularians (Fig. 10.1) are no more than a few millimeters long and have a tadpole-like shape. Although they appear very simple and minimally evolved, these plankton animals, along with other tunicates (such as salps, doliolids, and pyrosomes), are among the chordates, which are closest to us evolutionarily than any other planktonic organism (except for fish larvae, of course). They are chordates because they possess a notochord or nerve dorsal cord. With a bit of imagination, one could find some resemblance to the early stages of an embryo. However, the most fascinating aspect of these creatures is not this fact, but the house in which they reside. Several times a day, appendicularians construct a gelatinous house that serves as a highly efficient filtration device, allowing them to filter particles as small as thousandths of a millimeter. These houses are usually less than a centimeter long, and the appendicularians that inhabit them create water currents through rhythmic tail beats to swim and attract food (small protists and bacteria). With this system, they can filter a few liters of water per day.

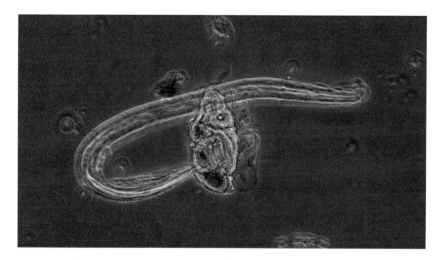

Fig. 10.1   Appendicularian. © Albert Calbet

Salps and Doliolids

Although salps and doliolids (Fig. 10.2) are distinct groups, they share many characteristics, so I will explain them together. Both are tunicates, like appendicularians, and are filter-feeders. They resemble barrels and are often mistaken for pieces of transparent plastic. Their length ranges from a few millimeters to a few centimeters. While they are frequently solitary, they can also form colonies that extend several meters in length. They inhabit virtually all seas and oceans, but in Antarctica, salps, in particular, play a crucial role in the ecosystem by consuming and packaging the microalgae that proliferate in the spring. In the Southern Ocean, they hold significant importance alongside the famous krill, to the extent that we refer to "years of salps" or "years of krill."

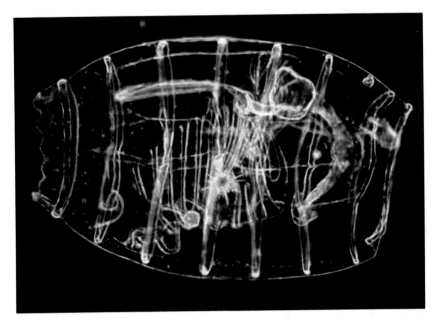

Fig. 10.2 *Doliolum nationalis.* © Albert Calbet

Pyrosomes

Pyrosomes are one of the rarest and most challenging groups to find within gelatinous plankton. They are also tunicates and consist of colonies of clonal organisms that are a few millimeters in size. Joined together by a gelatinous matrix, they form structures ranging from a few centimeters to about 20 meters in length, always in the shape of a wind sleeve. All individuals in the colony collaborate to swim and search for food. We can find them in the surface layers of tropical seas, as well as at great depths. As if these characteristics were not enough, they are also bioluminescent.

Jellyfish

Everyone is familiar with jellyfish. However, not everyone may know that they are close relatives of corals. In fact, many species undergo a polyp phase (similar to corals) during their life cycle, in which they are sessile.

From the polyp, small jellyfish (ephyra; Fig. 10.3) emerge through a process called strobilization. There are many types, shapes, and colors of jellyfish. According to taxonomy, there are four classes: schyphozoa, cubozoa, hydrozoa, and staurozoa. The schyphozoa are the true jellyfish, with their umbrella-shaped body (known as the medusa) and tentacles. Many of them have stinging cells called cnidocytes, which can deliver a painful sting. It is important to note that even if a jellyfish is dead on the beach (Fig. 10.4), its cnidocytes could still be functional, so it is best to avoid touching them. The cubozoa are cubic in shape, hence their name, and despite their small size, many Australian box jellyfish can be highly dangerous and even lethal. While a species of box jellyfish has been identified in the Mediterranean, it is not as dangerous as those found in Australia. Hydrozoans are colonial organisms, and examples include the Portuguese man of war or tiny blue *Velella* that often invade beaches and coastal areas. Within the colony, individuals assume different roles, with some acquiring buoyancy, others fulfilling reproductive functions, and

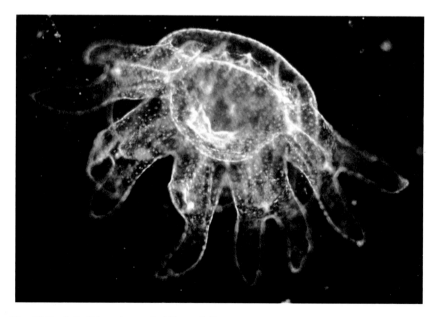

Fig. 10.3   Jellyfish ephyra. © Albert Calbet

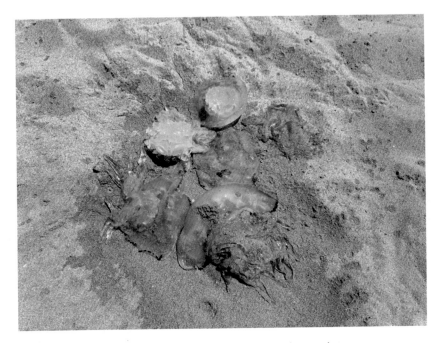

Fig. 10.4 Jellyfish (*Rhizostoma pulmo*) let to die on the sand

others hunting and digesting prey. Finally, we have the staurozoa, which are not very common, small in size, and sessile.

In places, like the seas of Japan, problematic species have proliferated, causing significant impacts on fishing activities. The Nomura jellyfish, which have seen a recent increase in abundance in these waters, can grow to more than two meters in diameter. Their high numbers often lead to the entanglement and destruction of fishing nets, resulting in the entire catch being thrown back into the sea. Moreover, they compete with fish for food (zooplankton) when they are small and even prey on fish as they grow larger. It is not a favorable situation, to say the least.

Ctenophores

Ctenophores (Fig. 10.5) are similar to jellyfish, but they have characteristics that set them apart. For starters, they do not sting (or not as much), and their locomotion is facilitated by eight bands of ciliated combs that beat together

Fig. 10.5 Ctenophore. © Albert Calbet

to generate movement. They come in various shapes, with a predominant spherical shape with or without tentacles, or a ribbon-like form. One of the best-known species is *Mnemiopsis leidyi*, an invasive species that has caused significant disruptions to fisheries wherever it has been introduced. The most notable example is the Black Sea, where this species was accidentally introduced in the 1980s. Within a decade, *M. leidyi* reached densities of about 400 individuals per cubic meter and decimated local fish populations. To mitigate this ecological and economic disaster, another ctenophore species, *Beroe ovata*, was introduced, as it preys on *M. leidyi*. This intervention seems to be effective, and while *M. leidyi* has not been completely eradicated from the Black Sea, its abundance seems now under control.

After reviewing these diverse creatures, one might question the usefulness of grouping them together as "gelatinous plankton." Each group we have discussed has distinct life cycles, feeding habits, evolutionary histories, and ecological strategies, making this classification somewhat arbitrary. Nevertheless, they all share a resemblance to jelly desserts.

Phronima: A Plankton Organism That Came to Hollywood

Tightly related to jellyfish we have some of their predators. The case of *Phronima* (Fig. 10.6) is particularly interesting. Belonging to the order Amphipods, *Phronima* leads a rather fascinating life. Not only are they famous, but their story is worth exploring. *Phronima* typically measures no more than four or five centimeters (a couple of inches) in length. They possess long legs, thick compound eyes, and are semi-transparent, adorned with red spots—characteristics that are fairly common among crustaceans. However, what sets *Phronima* apart is its unique life strategy. They swim through the depths of the ocean until they encounter a salp, a doliolid, or any other gelatinous planktonic tunicate. Once they find a suitable host, *Phronima* moves in, much like someone occupying a summer house.

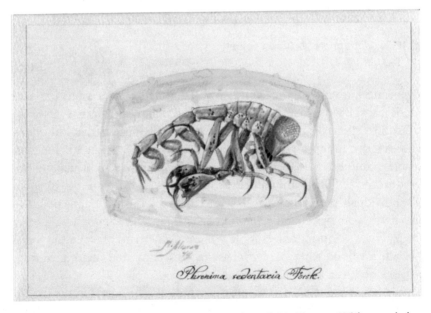

Fig. 10.6 *Phronima*. Drawing by Miquel Alcaraz. © M. Alcaraz. With permission

Equipped with formidable claws, *Phronima* carves out the insides of their hosts, leaving behind an empty barrel-like structure. Although the final form bears little resemblance to the original host, some cells within the structure remain alive. From the inside, *Phronima* navigates the sea, feeding on suitable prey. The transparent structure serves as protection and a place for a female *Phronima* to lay nearly 600 eggs. The eggs hatch within this "nursery," and the young develop inside until they reach a pre-mature stage, at which point they leave the structure and venture out into the ocean to start their own lives.

Now, you might be wondering why I mentioned that *Phronima* is famous. Well, it is believed that the terrifying creature that decimated almost the entire crew of the Nostromo spaceship (except for Lieutenant Ellen Ripley and her cat, Jonesy) in Ridley Scott's film "Alien" (1979) was inspired by *Phronima*. If you observe closely, you will notice the similarities.

11

Holoplankton and Meroplankton: Two Peculiar Terms for Common Creatures

A. Calbet, *The Wonders of Marine Plankton*,
https://doi.org/10.1007/978-3-031-50766-3_11

The group of zooplankton known as holoplankton spend their entire life cycle as plankton; this includes copepods, cladocerans, and protozoa. Meroplankton, on the other hand, spend only a part of their life as plankton, including larvae of starfish, sea urchins, crabs, and fish. Some meroplankton larvae have distinct forms from their adult counterparts, while others closely resemble them. For instance, flatfish larvae have eyes on both sides for a 3D view, adapting to planktonic life before developing into benthic adults.

Biologists often rely on Greek or Latin roots when naming organisms. This tendency has given rise to some curious and unusual terms, such as the ones we will discuss today: holoplankton and meroplankton. Both "holo" and "mero" come from Greek. "Holo" means "complete," while "mero" means "part of." This does not imply that some are whole organisms while others are mere fractions. Rather, it signifies that holoplankton spend their entire life cycle as plankton, whereas meroplankton spend only a portion of their life in this form.

We are already familiar with holoplankton (Fig. 11.1) from previous chapters: copepods, cladocerans, various protozoa, salps, and more. However, meroplankton (Fig. 11.2) may sound less familiar. But in reality, we come across numerous meroplankton organisms at the fish market or even during a trip to the beach. Starfish, sea urchins, crabs, and sea snails all have larval stages that exist as part of the plankton community. Some of these larvae exhibit bizarre forms that bear little resemblance to their adult counterparts.

On the other hand, some larvae closely resemble their adult forms. For instance, many fish species spend a portion of their early lives as planktonic larvae. Many jellyfish reproduce through a process called strobilation, transitioning from benthic polyp stages to free-swimming larvae. Even certain protists, like the dinoflagellates responsible for harmful algal blooms (commonly known as red tides), can encyst and spend a significant part of their life cycle in sediment, away from the planktonic realm. An intriguing example is the flatfish, such as turbot and sole. This fish typically resides on the ocean floor, and as a result, it has undergone a physiological adaptation: both of its eyes are located on the same side of its face. This arrangement allows it to have a three-dimensional

Fig. 11.1 The copepods are members of the holoplankton. *Centropages violaceus.* © Albert Calbet

Fig. 11.2 Members of meroplankton. Left, a larval stage of a brittle star. Right, a crab larva. © Albert Calbet

perception of everything situated above it. Also, I suppose it would be rather inconvenient to have an eye facing the bottom and constantly filled with sand. However, during its larval stages (Fig. 11.3), the sole has one eye on each side of its face. This is because the larvae are planktonic and require a three-dimensional view of what lies ahead and on their

Fig. 11.3 The image depicts the progression of eye migration across different larval stages of different species of flat fish. © Albert Calbet

sides. As the larva develops, one eye migrates from one side of the face to the other. Once both eyes are on the same side of the face, even though the fish is still tiny, it adopts a benthic behavior.

As you can see, meroplankton organisms are more numerous than we might think, and many of them are well-known. Holoplankton, too, encompass a wide range of organisms. Let's not forget that copepods are likely the most abundant multicellular animals on our planet, and the realm of holoplankton also includes viruses, bacteria, and protists.

12

Feeding Mechanisms in Zooplankton

Zooplankton exhibit a fascinating array of feeding strategies that are influenced by their environment and Reynolds numbers. From the viscous realm of protozoa to the inertial world of jellyfish, copepods, and others, their tactics vary widely. These tiny organisms must locate and consume food in highly diluted environments, compelling them to adopt ingenious techniques. These include passive and active ambush feeding, filter-feeding, and cruising feeding. These diverse strategies demonstrate the remarkable adaptability and convergence of zooplankton as they navigate their complex ecosystem in pursuit of survival.

Strange Numbers and Viscous Environments

Who can forget the iconic 1966 movie "Fantastic Voyage," featuring Raquel Welch and Stephen Boyd? The storyline revolves around an Eastern scientist who falls into a coma after an attack. The only way to save this character, who harbors highly significant secrets in his mind, is to miniaturize a submarine to the size of a blood cell and enter his body to reach the brain and remove a life-threatening clot. You may recall that the submarine in the movie used a propeller for propulsion, as is common in water vessels. However, a submarine of such small dimensions would make virtually no progress in water using a propeller. Why? Because of a magical number known as the Reynolds number. No, I have not gone crazy. The Reynolds number is a mathematical concept that distinguishes between fluids with laminar and turbulent flows. In simpler terms, it describes the interplay between viscous and inertial forces. Tiny particles have low Reynolds numbers and exist in the viscous realm, where water behaves like honey or syrup. Conversely, large particles have high Reynolds numbers, representing the inertial realm we are familiar with, where water flows freely through our fingers. What are the implications of this? The submarine in "Fantastic Voyage" (with a low Reynolds number) would struggle to move forward using a propeller because it would spin around without making any progress. To move effectively, it would require alternative methods such as a long flagellum or thousands of cilia. Within the realm of zooplankton, we encounter organisms that inhabit the viscous world (e.g., protozoa) as well as those in the inertial world (salps, jellyfish, etc.), and some that straddle both worlds, like copepods. The choice between these two realms influences their feeding strategies.

Prey, Where Are You?

Before delving into the details of zooplankton feeding, it is essential to note a shared characteristic among all zooplankton members—they must acquire their food from a highly diluted environment. For instance, to consume a ciliate (which is roughly the size of an apple to us) present at an abundance of less than one individual per milliliter, a copepod would need to explore at least 1000 mm^3. It might not seem like much, but considering that a copepod is only a millimeter long, it implies scanning and traversing a distance roughly ten times its length in every direction. Imagine having to search within a sphere with a diameter of 10–20 m to find and consume an apple. If you had to do it blindfolded (as copepods do not use sight to locate prey), you would likely starve. Fortunately, copepods and other zooplankton members possess various mechanisms to detect and attract prey.

Feeding Mechanisms in Zooplankton

Feeding by Diffusion of Inorganic or Organic Solutes Many unicellular zooplankton, particularly mixotrophs, can feed like algae by absorbing dissolved nutrients into their cells. Diffusion feeding is often accompanied by motility to disrupt gradients and prevent solutes from depleting near the cell. Some organisms, like *Karlodinium veneficum*, can simultaneously incorporate solutes through their membranes and consume prey. Passive Ambush Feeding Jellyfish, ctenophores, and certain protozoa like radiolarians and foraminifera drift and await prey. These organisms employ toxic structures to immobilize their prey, as well as other capture mechanisms, such as sticky pseudopods that aid in attracting the prey toward their mouth or vacuole. A noteworthy case is that of certain pteropods, tiny planktonic snails that secrete adhesive mucus nets similar to spiders to capture prey. Active Ambush Feeding Certain ciliated species like *Didinium* sp. or *Mesodinium rubrum*, as well as many copepods such as the ubiquitous *Oithona* spp. (Fig. 12.1) and chaetognaths, employ a capture mechanism akin to large felines in the savannah or jungle. They patiently wait for prey and pounce when they detect it. However, unlike lions that rely on sight to locate prey, zooplankton species have mechanosensitive structures that guide them toward their target. They also utilize specific chemical receptors, similar to the olfactory receptors found in terrestrial predators, to orient themselves toward the most favorable direction for finding prey. Filter-Feeders and Cruisers Appendicularians, salps, doliolids, cladocerans, and certain choanoflagellates utilize water currents to attract small prey such as bacteria and tiny phytoplankton to their efficient filter structures (Fig. 12.2). Many crustaceans like barnacles, some calanoid copepods, and krill also generate feeding currents to attract prey, without necessitating movement of the organism itself. In contrast, other copepods and protozoa create feeding currents while

Fig. 12.1 *Oithona davisae* is an ambush feeder. © Albert Calbet

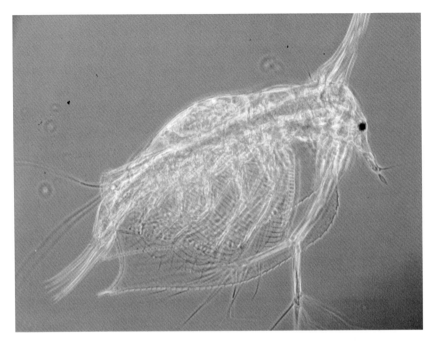

Fig. 12.2 *Penilia avirostris* is a filter-feeder cladoceran. © Albert Calbet

simultaneously using them for locomotion (cruising feeding). It is worth noting that copepods were previously believed to be filter-feeders until the early 1980s when it was discovered that they are suspensivorous and actively manipulate prey, often capturing and manipulating them individually or in small groups. Dr. Alcaraz and collaborators were the first to describe this mechanism, aided by movies showcasing the process using a *Eucalanus crassus* glued to a dog's hair with Crazy Glue.

As you can see, zooplankton employ a diverse range of methods to detect and capture prey, and these strategies are not exclusive to any particular group. There are frequent instances of evolutionary convergence toward different mechanisms.

13

Evading Predators: The Clever Tactics of Plankton Survival

In the vast expanse of Earth's oceans, a perpetual struggle unfolds between planktonic organisms and their predators. These microscopic beings are far from helpless; instead, they have refined a remarkable repertoire of survival strategies to navigate the perilous waters of predation. Among the most captivating and effective of these tactics are camouflage, mimicry, and defense mechanisms, each of which enables plankton to outwit detection, confuse predators, and mount a resolute defense against their relentless adversaries.

Evasive Mechanisms: Avoiding Unwanted Contact

Transparency At its core, camouflage is an ancient survival technique that revolves around the concept of becoming one with the environment. In the open ocean, where the vast distances between predators and prey make detection a critical challenge, mastery of disguise becomes the linchpin that separates life from death for planktonic organisms. One of the most widespread manifestations of camouflage in planktonic organisms is transparency. A multitude of species have evolved to become nearly translucent, rendering them virtually invisible to the eyes of predators (Fig. 13.1). This transparency goes beyond mere absence of color; it involves a sophisticated adaptation that includes light manipulation. By allowing light to pass through their bodies without scattering, these organisms effectively deceive the visual cues predators rely upon. Consider, for instance, siphonophores (Fig. 13.2) and comb jellies. These gelatinous beings possess transparent bodies that seem to dissolve into the aquatic surroundings, making them appear ethereal and otherworldly. Yet, their journey of adaptation does not stop at transparency alone. Comb jellies have evolved specialized iridescent cells capable of scattering light in a mesmerizing array of directions. This interplay between transparency and light manipulation stands as evidence to the ingenious ways plankton have harnessed their environment to enhance their chances of survival.

Mimicry Mimicry, a cryptic strategy, occurs when one species mimics another to gain protection. Planktonic organisms have woven mimicry into their defense mechanisms. In some cases, species that are toxic or unpalatable adopt the appearance of their harmful counterparts, effectively

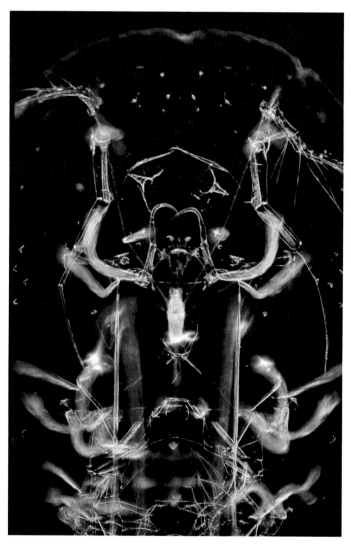

Fig. 13.1 Transparent copepod. *Sapphirina* sp. © Albert Calbet

deterring predators. For instance, some fish larvae mimic ctenophores and jellyfish. A very particular case, result of casualty, is that of jellyfish in the nowadays waters. These transparent creatures possess bell-shaped bodies reminiscent of drifting plastic bags. This (non-intentioned) resemblance misleads predators into thinking they are inedible, providing an escape route.

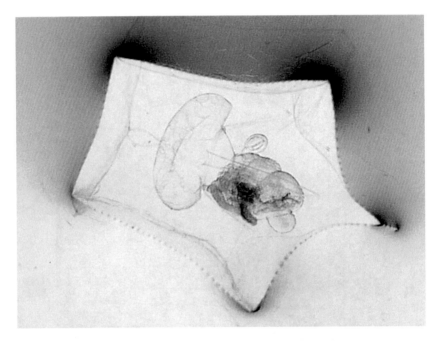

Fig. 13.2 Siphonophors are nearly transparent to the eyes of predators. © Albert Calbet

Defense Mechanisms: When the Threat Is Unavoidable

Moving Away from Predators Many plankton species exhibit escape behaviors such as rapid jumps, zig-zag swimming, and tumbling. These erratic movements can confuse predators and create difficulties in tracking the prey. Diel vertical migration, a behavior in which plankton move closer to the surface during the night to feed and retreat to deeper waters during the day to avoid visual predators, also enhances their survival.

Bioluminescence Bioluminescence takes a prominent spot in the planktonic arsenal and, as such, it will be discussed in the following chapter. However, here, I present a glimpse of its relevance. When threatened, some plankton deploy bioluminescent bursts. This sudden flash confounds or astonishes predators, offering the plankton a precious escape window. The interplay between predators and prey orchestrated by bioluminescence is a tale of complexity. Some predators exploit these luminescent bursts, using them as breadcrumbs to attract their victims. This dynamic has spurred an evolutionary arms race between plankton and predators, an ongoing struggle for dominance through adaptability.

Spines, Armor, and the Fortifications of Defense   In addition to evasive tactics, certain plankton have developed physical adaptations that discourage predators. Some species possess spines (Fig. 13.3), barbs, or armored structures (Fig. 13.4), making them less appealing or challenging to consume. These adaptations often complement rapid movements

Fig. 13.3   The spines of acantharids are a very efficient defense against grazers. © Albert Calbet

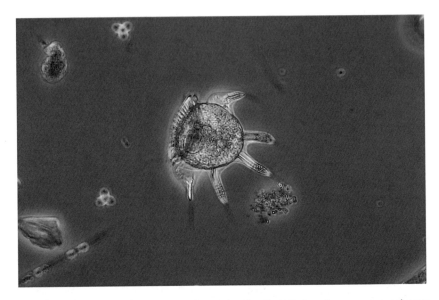

Fig. 13.4 The armored structures of certain dinoflagellates discourage predators. © Albert Calbet

or the capacity to retract appendages, reinforcing their ability to evade predator attacks. Planktonic copepods, crustacean larvae, rotifers, protists, etc. possess spines and setae that render them less appetizing. These projections, besides fending off attacks, discourage predators from engaging. Similarly, specific dinoflagellates have crafted external plates, providing an additional layer of protection against grazers.

Toxicity and the Shield of Chemical Defense Plankton's realm teems with organisms that have refined the art of chemical defense. They concoct compounds that deter predators upon ingestion or emit chemicals into the water, creating an unwelcoming environment. Toxic algae are a prominent example of chemical warfare. Among them, dinoflagellates stand out. They manufacture toxins that carry profound consequences for predators. Consuming these toxins leads to poisoning, paralysis, or even death. This intricate interplay sends ripples through the food web, as predators learn to avoid these hazardous morsels, adapting to alternative sources of sustenance.

Evolutionary Arms Race: A Dynamic Duel

The relationship between predators and plankton is often compared to an evolutionary arms race. As predators develop new ways to hunt and capture their prey, plankton respond with equally ingenious defenses. This dance of adaptation has shaped the course of evolution, fostering the development of remarkable survival strategies on both sides.

Consider the ever-elusive copepods once again. These tiny creatures have evolved to have complex escape mechanisms that allow them to twist and turn, making it extremely difficult for predators to catch them. In response, predators have developed specialized behaviors and structures to better capture copepods. This ongoing battle of wits drives the continuous refinement of adaptations, painting a vivid picture of nature's constant quest for equilibrium.

14

Bioluminescent Plankton

Bioluminescent beaches are captivating spectacles caused by marine creatures emitting glimmering blue or green lights. While relatively rare along coastlines, bioluminescence is well-studied and common among various organisms. It is observed in bacteria, protists, algae, jellyfish, ctenophores, squid, crustaceans, and fish. The primary purpose of this light emission is to deter predation by confusing or attracting predators. Even marine bacteria use bioluminescence to communicate, repel predators, and thrive. The process involves enzymes, luciferase, and luciferin, which generate light energy through a chemical reaction. Understanding the origin, causes, and mechanisms of bioluminescence unveils the wonders of nature's light shows.

Those fortunate enough to witness a bioluminescent beach describe it as a breathtaking spectacle, unlike anything else on Earth. The glimmering blue or green lights dancing with the breaking waves captivate both young and old alike. Unfortunately, I have not had the chance to witness it myself, although I have seen occasional sparks in the nighttime waters and have had the joy of stirring a culture of bioluminescent protozoa.

Despite the infrequent occurrence of bioluminescence along the coast, limited to specific beaches and sporadic events, it is a well-studied and relatively common phenomenon among marine creatures. Even the U.S. military has explored its potential for detecting enemy submarines. However, what interests us here is not the military applications of this natural curiosity but rather its origin and causes.

What Organisms Emit Light?

There are various types of bioluminescent organisms—both marine and terrestrial—including bacteria, protists, jellyfish, ctenophores, squid, worms, mollusks, crustaceans, echinoderms, and fish, among others. However, our focus lies on those belonging to the planktonic realm. Specifically, we are interested in the majority of unicellular protozoa (Fig. 14.1) and algae (especially dinoflagellates), some jellyfish and ctenophores, as well as certain crustaceans, predominantly from the copepod group.

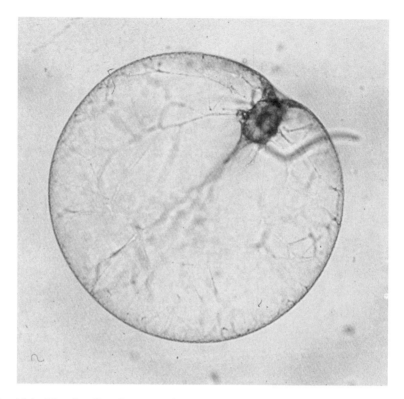

Fig. 14.1   The dinoflagellate *Noctiluca scintillans* is responsible for many bioluminescent displays worldwide. © Albert Calbet

Why Do They Glow?

Terrestrial groups, such as insects, use luminescence to attract mates.
However, for plankton, the primary purpose of bioluminescence is to
deter predation. Imagine you are a copepod suddenly sensing the suction
from the mouth of a fish. While you may have the ability to escape and
propel yourself at high speed, if you can also confuse your predator by
emitting a flash of light, rendering it unable to track you, your chances of
successful escape are significantly enhanced (Fig. 14.2). Additionally,
there are abyssal fish that emit light, luring curious prey toward them,
acting as fools drawn to the light and ultimately ending up in the mouth

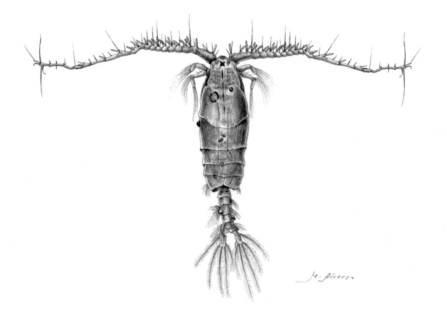

Fig. 14.2 The copepod *Pleuromamma abdominalis* emits flashes of light when it senses danger. Drawing by Miquel Alcaraz. © M. Alcaraz. With permission

of the predator. In this case, the light originates from symbiotic bacteria residing in various cavities of the fish.

Marine bioluminescent bacteria present an intriguing case. Many bioluminescent bacterial groups are symbionts of higher organisms, like cephalopods or fish, where they serve functions, such as attracting prey (as previously mentioned), repelling predators, or even communication. However, there are also free-living bioluminescent bacteria. Until recently, it was believed that bioluminescence was associated with high population densities of these organisms, serving as a response to such conditions. However, recent studies have shown that bioluminescence can occur in individual bacteria. In such cases, it is presumed that emitting light is a mechanism to attract predators, especially in environments where nutrients are scarce. When ingested, these bacteria, which are also resistant to digestion, find a richer environment within their host, thereby increasing their chances of survival.

How Do They Produce Light?

Bioluminescence occurs when the enzyme luciferase reacts with oxygen and a protein called luciferin. The process unfolds as follows: Oxygen, with the assistance of luciferase, oxidizes luciferin, and utilizing the energy stored in ATP, generates light energy along with water as a by-product. It may sound complex, but even a bacterium is capable of performing this remarkable feat!

Bioluminescent plankton are a fascinating phenomenon that showcases the marvels of nature. By understanding the organisms that emit light, the reasons behind their illumination, and the mechanisms involved, we gain insight into this captivating natural occurrence.

15

Global Warming and Plankton

The Earth's accelerated warming carries dire consequences, including extreme weather, sea-level rise, and shifts in species distribution. Temperature impacts plankton greatly: ectotherms' metabolic processes quicken, leading to physiological changes and potential imbalances. As species adapt, some thrive while others face extinction due to extreme temperatures and changing ecosystems. Oceans' vital role in climate regulation and ecosystem dynamics underscores the urgency for CO_2 emission reduction. Without action, warming could reshape marine ecosystems, favoring smaller, warmer-climate organisms and endangering food webs and fisheries. Conservation is key to securing a sustainable future for generations to come.

Today, there is no denying that the Earth is warming at a faster rate than what would be expected from natural causes alone. From a human perspective, one might think that a couple of degrees of temperature increase Will not make much of a difference. Some may even view global warming as a positive thing, assuming it will bring milder winters and slightly warmer summers, especially when we can rely on air conditioning. However, this could not be further from the truth. Even a small rise in average temperature can have catastrophic consequences for our current way of life. These consequences include extreme droughts, heat waves, more frequent and intense storms, heavy rainfall, sea-level rise, shifts in species distribution, mass die-offs of animals and plants (remember that plants cannot escape), and more.

Temperature Effects on Plankton

For ectotherms (animals whose regulation of body temperature depends on external sources), as the temperature rises, metabolic processes accelerate. Scientists measure the effect of temperature on various metabolic rates using the Q_{10} coefficient (not to be confused with the coenzyme of the same name). The Q_{10} value represents the increase in a process when the temperature rises by 10 °C. Then, a Q_{10} of 2 implies that a metabolic process doubles in speed when the temperature increases by 10 °C. It is important to note that not all rates respond equally to temperature, and abrupt increases can disrupt different functions. However, given sufficient time and a gradual temperature change, within a thermal range, physiological rates tend to balance out through a process called thermal acclimation. When this response includes changes in the genome of the organism transmitted across generations, we refer to it as thermal adaptation.

Before reaching equilibrium, organisms experience metabolic uncoupling, which can reduce their survival capacity and potentially lead to death. Therefore, when discussing the resistance of an organism to temperature, we typically define its thermal window (Fig. 15.1), which varies for each species adapted to its specific environment. If a species inhabits both tropical and cold zones, individuals in tropical regions usually exhibit values shifted to the right, reflecting higher values at higher temperatures. The plankton community encompasses a diverse range of thermal amplitudes, spanning from polar and tropical regions to hydrothermal vents and tide pools that can experience significant temperature fluctuations within 24 h. Regardless of the harshness of the conditions, there will always be species that find their niche and survive.

While species generally adapt to their current thermal conditions, they also exhibit plasticity and can potentially invade and fully adapt to new habitats over time. In laboratory settings, it has been demonstrated that

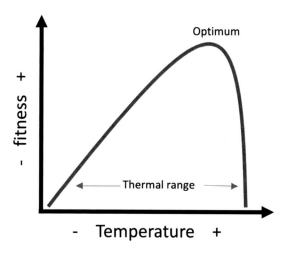

Fig. 15.1 Example of thermal window. © Albert Calbet

both algae and copepods (and likely protozoa as well) can regulate their metabolic rates and counteract the effects of temperature after a prolonged period (around a year) of genetic adaptation to higher temperatures. Initially, when exposed to a temperature 5 °C above their normal habitat, the respiration rate of an alga surpasses its photosynthesis rate. This is because respiration is more sensitive to thermal changes compared to photosynthesis. However, over many generations in the new temperature conditions, the two rates gradually rebalance themselves. Now, you may wonder why this is a concern. The issue is that during this adaptation process, which can last months or even years, the species experience metabolic imbalances and become less competitive compared to other better-adapted species. A clear example of this is the displacement of the cold-water copepod *Calanus finmarchicus* by the warmer-water copepod *Calanus helgolandicus* in the North Sea. While *C. helgolandicus* reproduces at a slower rate, it cannot sustain cod populations as effectively as *C. finmarchicus*, which is highly nutritious and prolific. Consequently, the collapse of important fisheries relying on *C. finmarchicus* may be imminent.

The critical issue regarding temperature rise also lies in the extremes: species from very cold regions (Fig. 15.2) will inevitably vanish if temperatures continue to rise, while those from warm climates will face

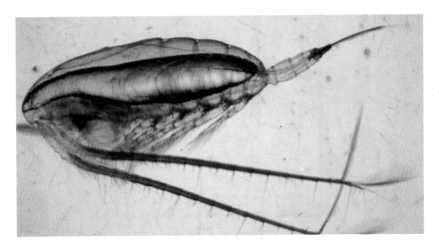

Fig. 15.2 *Calanus hyperboreus* is a polar copepod species threatened by warming. © Albert Calbet

extreme challenges and risk extinction if they cannot adjust their metabolism in time. Heatwaves, which are increasingly frequent, pose serious difficulties for many organisms within a matter of days.

The Role of the Ocean on the Climate

Earth's climate cannot be comprehended without considering the role of the oceans, and vice versa. Water has a much higher heat capacity than soil, which allows it to store and release heat more slowly. This leads to phenomena such as daily sea breezes and other climatic patterns. Temperature differences between polar regions and the equator drive a planetary circulation of water, transferring heat (energy) and nutrients around the planet through ocean currents. However, variations in temperature and atmospheric circulation can affect this transport. For instance, the periodic melting of the Arctic ice sheet in spring causes the resulting cold water to sink near the east coast of Greenland, generating a displacement of water masses that drives the Gulf Stream and influences the entire global oceanic circulation (often referred to as the conveyor belt). In the event of a significant reduction of the surface area of the northern polar ice sheet, the associated cold water from spring melting may be insufficient to activate the Gulf Stream. The climatic consequences of such a phenomenon are highly uncertain, although they may not reach the extremes portrayed in the movie "The Day After Tomorrow" where the United States freezes within hours. Nevertheless, significant weather changes would undoubtedly occur.

Climate change can also impact the timing and intensity of upwellings, which could have profound implications for major fisheries and overall productivity within the marine ecosystem. Warming of the oceans can alter the direction and strength of currents and disrupt the distribution of marine species. Phenomena like "El Niño" are directly influenced by climatic conditions and control the flow of nutrients, which in turn affects fisheries along the west coast of South America. On a more local scale, the introduction of new species (sometimes facilitated by the transport of organisms in ship ballast water or through species exchange in aquaculture), the increased frequency and severity of harmful algal blooms (also related to other human impacts), and the expansion of anoxic zones in seas and oceans are just a few examples of the changes awaiting us in the immediate future.

Is There a Solution?

The short answer is no. However, what we can do is slow down the rate of temperature increase. In the 2016 Paris Agreement, signed by the majority of United Nations countries (unfortunately, some of the most polluting nations refused to sign), participating countries pledged to reduce CO_2 emissions by a certain percentage as part of the effort to mitigate global warming. The intention behind this commitment was to mitigate the rise in global temperature. Regrettably, it appears that only a few countries are actually following through on their promises, and the temperature continues to rise. To put it into perspective, during the first wave of the SARS-CoV-2 pandemic in 2020, when economic activities, transportation, and industrial production were significantly reduced for 2 months, we only achieved the annual emission reduction required by the Paris Agreement. Moreover, significant results from reducing CO_2 emissions may not be visible for another 50 or 60 years.

So, What Does the Future Hold?

Regarding plankton, we are certain to witness changes in the coming years, as already observed. Some experts predict a significant dominance of jellyfish and other gelatinous plankton in most oceans, along with a substantial decrease in algae production and its cascading effects throughout the entire food web. It is important to consider that organisms from warmer climates are generally smaller than their counterparts from colder regions, leading to shifts in the structure of food webs and ultimately impacting the entire marine ecosystem. A mere few degrees of average water temperature can disrupt the upwelling of cold, nutrient-rich waters, alter currents (thus affecting the climate of the planet), and trigger massive migrations in marine organisms, although not all species have the means to escape such changes. Unfortunately, most scenarios point to a reduction in fish populations.

It is evident that the conservation of ecosystems has never been a strong incentive for governments, perhaps due to a lack of education and awareness regarding the importance of nature. Nature is far more complex and interconnected than we often perceive. If one link fails, others may collapse. It is crucial that we all comprehend the significance of conserving nature as our only solution to ensure the survival of future generations. If conserving energy, reducing emissions, and taking necessary measures to protect four species that are at risk of disappearing may not seem relevant to you, consider doing it for your descendants. They may not have the ability to act now, but they will inherit our legacy. Alternatively, we can choose to do nothing and wait for the wave of mass extinction to engulf us. The choice is yours.

16

Plankton Adaptations to Extreme Environments

A. Calbet, *The Wonders of Marine Plankton*,
https://doi.org/10.1007/978-3-031-50766-3_16

In the ever-changing landscape of aquatic environments, planktonic organisms have proven themselves to be adaptable and resilient. Nowhere is this adaptability more apparent than in extreme environments, where these microscopic life forms have developed ingenious strategies to endure conditions that challenge the limits of life as we know it. From the abyssal depths of hydrothermal vents to the frigid expanses of polar waters, plankton have harnessed the power of evolution to thrive against all odds.

Deep-Sea Hydrothermal Vents

Deep beneath the waves, where sunlight struggles to penetrate, lies a realm that defies expectations—hydrothermal vents. These underwater volcanoes spew forth scalding-hot water loaded with minerals from the Earth's mantle. In an environment where temperatures can exceed 350 °C and pressures reach crushing depths, one might assume life to be impossible. However, hydrothermal vents are bustling with activity, and planktonic organisms are at the heart of this challenging ecosystem.

Plankton (mostly bacteria and archaea) in these extreme environments have evolved adaptations that are nothing short of astonishing. The vent-dwellers are not dependent on sunlight, as their counterparts in more hospitable waters are. Instead, they harness the energy from chemical compounds dissolved in the vent fluids through a process known as chemosynthesis. This remarkable adaptation enables them to bypass the need for sunlight and photosynthesis.

To survive in these harsh conditions, planktonic organisms near hydrothermal vents have developed specialized enzymes and heat-resistant proteins that allow them to function at high temperatures. Additionally, their cell membranes are equipped with unique lipid compositions that enhance their resilience to pressure and temperature extremes. These adaptations have enabled the emergence of an ecosystem that thrives on a completely different energy source than the rest of the world's ecosystems.

Thriving in the Icy Realms

The polar regions, with their icy expanses and subzero temperatures, might seem inhospitable to life, yet planktonic organisms have carved out niches in these frozen realms. For instance, diatoms, a group of phytoplankton, are among the most successful inhabitants of polar waters. These microscopic algae have evolved an array of adaptations to cope with the unique challenges posed by the extreme conditions. One of the most fascinating adaptations is the production of antifreeze proteins. These proteins prevent the formation of ice crystals within the diatoms' cells, allowing them to continue photosynthesizing even when surrounded by freezing water. Interestingly, diatoms have an intricate cell structure that provides buoyancy. This enables them to stay suspended in the upper layers of polar oceans where sunlight is available, maximizing their photosynthetic potential.

Also, other plankters have efficiently colonized these waters, and proven unique adaptations to the freezing environment and extreme dichotomy of weather. Some copepods, for instance, adapted their life cycle to the seasonality of polar ecosystems, enduring sort of dormancy during winter. Winter krill obtains energy by feeding on the algae trapped in ice. These are merely examples of the intriguing adaptations plankton have developed through thousands of years.

Adaptations to Extreme Pressures: Prospering in the Deep

Beyond the reach of sunlight, the pressure of the water column increases exponentially. In the dark and crushing deep ocean, planktonic organisms have evolved adaptations that allow them to withstand the extreme pressures that would crush most organisms. Their cell membranes are designed to be flexible and durable, enabling them to survive under the immense weight of the water above. Some plankton, like certain copepods, can conduct daily migrations from surface up to 1000 m deep. They have specialized oil-filled sacs that provide buoyancy, allowing them to float effortlessly through the water column (Fig. 16.1). These adaptations not only help them conserve energy but also enable them to avoid expending unnecessary effort to counteract the downward pull of gravity.

Moreover, deep-sea bacteria exhibit remarkable swimming capabilities even under the extreme conditions of 150 MPa, indicating their unique adaptation to high-pressure environments. Additionally, certain bacteria have demonstrated the ability to generate a distinct set of flagella (known as lateral flagella) in response to elevated pressure or lowered temperatures.

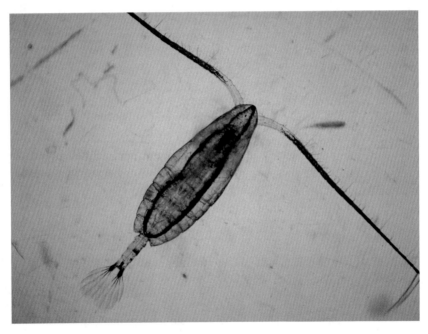

Fig. 16.1 *Calanus* sp. Copepod showing an evident oil sac in the center of the body. © Albert Calbet

Coping with High Solar Irradiation at the Surface

At the opposite end of the aquatic spectrum lies the surface of the water, where intense solar radiation bombards the upper layers. Here, planktonic organisms have developed adaptations to cope with the high-energy photons from the sun. Some phytoplankton have evolved pigments that absorb light efficiently while protecting their cells from the damaging effects of ultraviolet (UV) radiation. Additionally, certain planktonic organisms have developed the ability to adjust their position in the water column to regulate their exposure to sunlight. They migrate vertically, moving deeper into the water during the brightest parts of the day and ascending closer to the surface during periods of lower light intensity. This behavior helps them optimize their photosynthetic efficiency while minimizing the risk of UV damage.

In situations where vertical displacements are unattainable or when extremely shallow waters define the chosen habitat, species display photoprotective compounds such as mycosporine-like amino acids and carotenoids, alongside antioxidant capabilities and repair mechanisms. This phenomenon is evident in specific copepods (Fig. 16.2), certain siphonophores, and numerous other micro and macro-organisms inhabiting the ocean's uppermost layer, collectively referred to as the neuston.

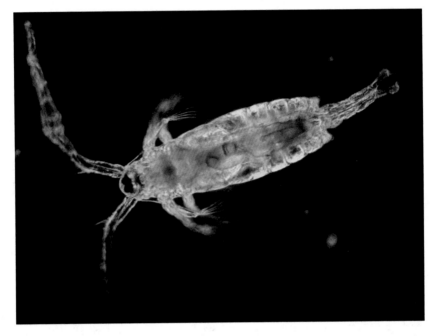

Fig. 16.2 *Labidocera wollastoni* is a neustonic copepod that displays a blue photoprotector pigment. © Albert Calbet

Living Without Oxygen

Planktonic organisms have developed remarkable adaptations to thrive in oxygen-depleted waters, demonstrating their resilience in challenging environments where oxygen availability is limited. These adaptations are crucial for their survival, given that oxygen is a fundamental component of energy production and cellular respiration. In response to these conditions, various planktonic groups have evolved specialized strategies to confront the lack of oxygen.

Certain species have embraced alternative metabolic pathways, such as anaerobic respiration, to sustain their energy needs. For instance, some bacteria within the Bacteroidetes group employ nitrate respiration when oxygen is scarce, enabling them to extract energy from organic matter. Similarly, sulfate-reducing bacteria, such as those from the Desulfobacteraceae family, utilize sulfate as an electron acceptor during anaerobic metabolism.

Planktonic diatoms and dinoflagellates showcase adaptations that enhance their oxygen uptake efficiency. Diatoms are particularly efficient at surviving under low oxygen conditions. They are even able to respire from nitrate when trapped in anoxygenic sediments. Also, dinoflagellates like *Oxyrrhis marina* can survive at very low oxygen concentrations. In response to oxygen-deficient conditions, some planktonic species can enter states of metabolic dormancy. For example, certain copepod species can exhibit diapause, a period of suspended development, to conserve energy until oxygen levels improve. This adaptation safeguards their survival during extended periods of oxygen depletion. Moreover, behavioral adaptations also play a vital role in the survival of plankton in oxygen-depleted zones. Zooplankton and many flagellated and ciliated protists engage in vertical migration to avoid low-oxygen layers.

In conclusion, planktonic organisms thriving in anoxigenic waters have harnessed diverse adaptations to combat the challenges posed by limited oxygen availability. These strategies encompass metabolic, physiological, and behavioral mechanisms, underscoring the remarkable resilience of planktonic life in these demanding aquatic ecosystems.

The Lessons of Extreme Adaptations

The adaptations of planktonic organisms to extreme environments offer valuable insights into the limits of life and the diverse ways in which life can persist and flourish. These adaptations showcase the power of evolution to shape organisms in response to the challenges posed by their surroundings. The resilience of hydrothermal vent organisms, polar diatoms, and deep-sea inhabitants challenges our understanding of habitability and expands the potential niches where life could exist beyond Earth.

The study of these adaptations also holds promise for applications in various fields. Enzymes and proteins evolved by vent-dwelling organisms, for instance, have potential industrial and biotechnological uses due to their stability in extreme conditions. Additionally, understanding the strategies that allow organisms to survive in these environments provides clues for engineering materials that can withstand high temperatures, pressures, and solar irradiation.

17

The Fragility of Plankton

Plankton are often overlooked when considering the impact of disasters like oil spills. Pollution and anthropogenic activities also harm plankton, disrupting the marine ecosystem's delicate balance. Examples of how even minor changes affect plankton's reproductive behavior and primary production emphasize their sensitivity. Plankton vulnerability to pollutants, from heavy metals to emerging substances, raises concerns about their future resilience in changing environments. Recognizing their significance is crucial for safeguarding marine ecosystems.

When a disaster strikes the sea, like the Exxon Valdez incident in Alaska (1989) or the Prestige oil spill on the NW Spanish coast (2002), our immediate concern turns to seabirds, turtles, and dolphins, as well as the impact on fisheries. We witness images on television of oil-soaked birds and dead fish littering beaches. Undoubtedly, these visuals are relevant and important. However, very few people (aside from a handful of specialized scientists) stop to consider the potential consequences for the foundation of the marine food web (Fig. 17.1): the plankton. We fail to recognize that if the plankton are affected, the entire ecosystem will crumble like a house of cards. In such a scenario, birds, turtles, dolphins, and fish would be left without sustenance, leading to their irreparable demise. Thankfully, certain groups of plankton exhibit resilience against hydrocarbon pollutants, benefiting from the dilution effect of seawater and the hydrocarbon mostly superficial distribution.

It does not require a catastrophe to harm plankton. Anthropogenic activities alone have a significant impact. Consider this: if someone proposed cutting down an oak forest to make way for planting exotic fruits like passion fruit or papaya, we would rightfully condemn it as an ecological monstrosity. Yet, if the same were suggested for introducing a foreign species of oyster (like the ones commonly found in aquaculture facilities and mostly originating from the Pacific) in a bay where swimming is uncommon, it would likely go unnoticed. However, such practices destroy planktonic diversity and disrupt the delicate balance of the marine ecosystem surrounding them.

These oyster crops not only filter out a substantial portion of the existing plankton but also pollute the water and sediments to the extent of inducing anoxia. Furthermore, they become hotspots for the proliferation of jellyfish

Fig. 17.1   Plankton constitute the base of the trophic food web in marine ecosystems. © Albert Calbet

and harmful algal blooms, ultimately resulting in a genuine ecological disaster that often goes completely unnoticed.

Some Examples

Despite their multiple adaptations to certain extreme environments, plankton are highly sensitive to pollutants, a fact well-documented by numerous studies. Instead of inundating you with technical details and data on the effects of mercury, cadmium, chromium, or the consequences of sunscreen or pharmaceutical residues ending up in the sea, I will share two anecdotal examples that I personally witnessed. Although these examples were not part of any specific study, they undeniably highlight the fragility of plankton.

The first example dates back to my time as a Ph.D. student when I conducted laboratory cultivation of the copepod *Paracartia grani* (Fig. 17.2). To maintain these small crustaceans, we filtered seawater through cartridges

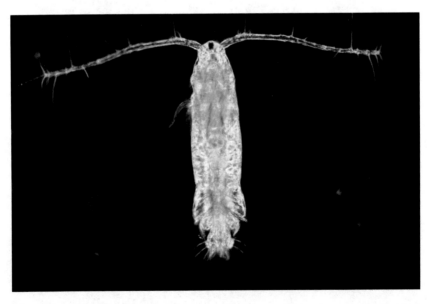

Fig. 17.2 Copepod *Paracartia grani* from the permanent culture at the Institute of Marine Sciences, CSIC, Barcelona, Spain. © Albert Calbet

fitted with different filters. I distinctly recall an incident when the plastic connecting screw between two cartridge holders got damaged, and we replaced it with a copper/bronze one. Almost immediately, despite appearing healthy, the copepods ceased to lay eggs. We meticulously checked their food, temperature, salinity, and other parameters but found no apparent issue. Finally, after exhausting all other possibilities, we realized that the only new variable had been the screw. Upon replacing it with a plastic one, the copepods resumed their usual egg-laying behavior. It is now known that copper inhibits vitellogenesis (egg yolk formation) in copepods, negatively impacting their reproductive output.

The second curious example occurred during my postdoctoral research at the University of Hawaii (USA). At the time, we conducted monthly oceanographic surveys in the Pacific Ocean, in which I had the privilege of participating. During these scientific cruises, we collected seawater samples using Niskin bottles (plastic cylinders with caps on both ends) arranged in a rosette configuration around a CTD (a device that measures conductivity, temperature, and depth; Fig. 17.3). On one occasion, the O-ring in two of the Niskin bottles, which ensures a proper seal and prevents water loss, broke. Since we did not have replacement O-rings of the same high-quality material (free from trace metals), the technician replaced them with standard black rubber O-rings commonly used in plumbing. Subsequently, in the following cruises, we consistently observed a decrease of approximately 20–30% in measurements of primary production (algae production) in those two bottles. Eventually, new O-rings of better quality arrived, and the issue was resolved. This seemingly insignificant detail, as the O-rings barely came into direct contact with the water, had a negative impact on the algae and forced us to discard all the data from those sampling events. It is worth noting that most Niskin bottles in use today still employ the standard black rubber O-rings. Since this finding was never published, it has not reached the wider scientific community. Moreover, publishing it would require further tests with different communities and increased replication. Nonetheless, the problem persists.

Numerous pollutants can affect plankton, including well-known ones such as hydrocarbons and heavy metals, as well as emerging pollutants that have received less attention, such as nanoparticles, hormones, medicines, nutrients, plasticizers, and more. We possess limited knowledge

Fig. 17.3 Rosette of Niskin bottles mounted around a CTD. © Albert Calbet

regarding the potential impact of these substances on plankton. Additionally, we remain uncertain about how rising temperatures may interact synergistically with these pollutants to the detriment of plankton. There is still much to learn. Meanwhile, I urge you to make a conscious effort to recognize that the planktonic food web, though invisible to the naked eye, constitutes an ecosystem as complex and invaluable as an oak grove or a pine forest.

18

From Land and Sea

The contrast between marine and terrestrial ecosystems shapes the distinct characteristics of their inhabitants. From physical environments to life strategies, differences arise due to factors like three-dimensional distribution in water columns and unique survival challenges. Marine food webs are intricate, driven by phytoplankton, while species succession and circadian migrations play vital roles. These variations, influenced by ocean currents and temperature dynamics, result in diverse marine organisms and ecosystems, showcasing the rich complexity of life below the waves.

Some cultures have the custom of combining ingredients from both mountain cuisine and the sea in their traditional dishes. While enjoying these dishes, few people consider the similarities and differences between marine and terrestrial organisms. Why are shrimp and chicken so distinct? Beyond the fact that some breathe through gills and others through lungs, their differences are deeply rooted in the foundations of their respective ecosystems.

The first characteristic that undoubtedly sets apart these ecosystems is the physical environment. Terrestrial ecosystems, such as forests, meadows, deserts, or tundra, are defined by a solid substrate and air as the primary medium for obtaining gases for breathing or photosynthesis. In certain marine ecosystems, like coral reefs, sandy bottoms, or seagrass meadows, there is also a solid substrate. However, the majority of living matter in the ocean resides in the water column. Life in the sea is predominantly distributed in three dimensions, whereas on land, its distribution is two-dimensional. Although valleys and mountains on land have a three-dimensional aspect, and trees reach toward the sky, most life occurs within a space that extends a few meters above and below the solid substrate.

The fact that marine organisms are distributed throughout the water column, a medium that is more viscous than air but not enough to ensure buoyancy without special mechanisms, makes them distinct from terrestrial organisms. It is also challenging to go unnoticed and hide in the water column. Perhaps the necessity to find efficient life strategies in such a unique environment has contributed to the exceptional biodiversity in the sea. Additionally, it is important to remember that life originated in the sea and later colonized land.

Essentially, in both land and sea, we find groups with similar functions but with different appearances and relative abundances. In both ecosystems, decomposers, producers, viruses, herbivores, omnivores, and carnivores can be found, but most of them bear little resemblance to their counterparts. For instance, primary producers on land are primarily plants and trees, while in the sea, this role is mostly fulfilled by microalgae (phytoplankton). Despite representing less than 1% of the biomass (weight) of terrestrial plants, phytoplankton globally produce the same amount of oxygen. The scarcity of nutrients in well-lit areas has led to the fact that most components of phytoplankton are also predators (mixotrophs). Phytoplankton find their main consumers in other unicellular organisms, called protozoa (or microzooplankton), and these, in turn, are consumed by larger metazoan organisms, such as copepods. Marine food webs are generally much more intertwined and complex than terrestrial ones, with numerous interconnections between groups at the same trophic level and with others from different trophic levels.

Another factor that distinguishes marine ecosystems from terrestrial ones is seasonal succession. On land, the transition between seasons results in changes in the landscape due to the response of trees and plants to climate and illumination variations, but essentially, the trees and plants remain the same throughout the year. Animals also tend to remain within the same eco-zone throughout the year, except for migratory animals such as certain birds and insects. In contrast, in the sea, modifications in primary producers and species succession throughout the seasons shape the entire ecosystem. These changes are elegantly represented in Professor Margalef's Mandala (1978; Fig. 18.1). This representation demonstrates how different species of marine phytoplankton appear throughout the year, associated with inorganic nutrients and the stability conditions (turbulence) of the water column. In general, when abundant nutrients and low turbulence are present, communities dominated by rapidly growing and often toxic dinoflagellates are expected. When nutrients and turbulence are high, diatoms dominate. Conversely, in nutrient-poor and low turbulence conditions, non-blooming dinoflagellates are present, along with small-sized cells that have large surface-to-volume ratios. Finally, under conditions of high turbulence and nutrient scarcity,

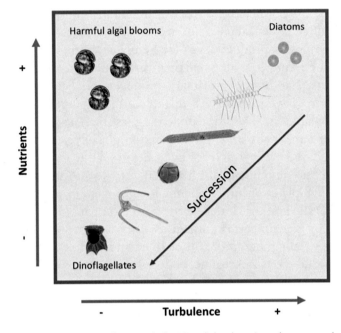

Fig. 18.1 Representation of Margalef's Mandala showing the succession of different phytoplankton species according to turbulence levels and nutrient concentrations. © Albert Calbet

phytoplankton are practically absent. Zooplankton follow this prey succession with more or less consistency year after year, also exhibiting changes in species composition (Fig. 18.2). For example, marine cladocerans, which feed on small prey, appear in the summer, while different species of copepods take center stage toward the end of spring and the beginning of winter.

It is also important to consider that ocean currents and water temperature influence this succession of species. These factors combined regulate the physicochemical dynamics of inorganic nutrients. Cold waters are usually associated with a higher concentration of nutrients and, therefore, promote the proliferation of phytoplankton, its predators, and ultimately, large-sized fish or other animals.

Another distinguishing feature is the day-night cycle. Like on land, the sea is home to diurnal and nocturnal animals; however, an

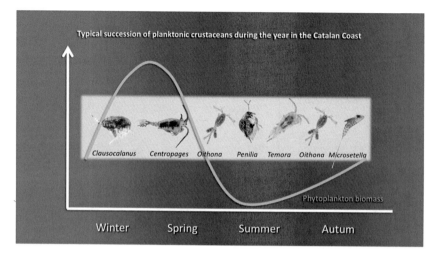

Fig. 18.2 The succession of the most abundant planktonic crustacean genera throughout the year on the Catalan coast, NW Mediterranean. The highest temperatures are indicated in red, and the lowest in blue. The abundance of phytoplankton is indicated by the green line. Data from Calbet et al. 2001. © Albert Calbet

impressive number of marine organisms undertake daily migrations to feed in shallower zones at night while spending the day at greater depths. On land, no population movement can compare to the circadian migrations of zooplankton.

When considering larger organisms, it is important to remember that water has a high specific heat. As a result, temperature changes are dampened in aquatic environments compared to terrestrial environments, although this depends on the size of the water mass. Therefore, marine organisms do not need to retain heat in their bodies, with the exceptions of marine mammals, which recolonized the marine environment from land, or some large predatory fish that generate heat.

As you can see, despite certain similarities, marine and terrestrial ecosystems are fundamentally different, leading to the presence of diverse organisms.

19

Plastic: You Are Plankton and To Plankton You Shall Return

A. Calbet, *The Wonders of Marine Plankton*,
https://doi.org/10.1007/978-3-031-50766-3_19

Plastic, with diverse origins, has revolutionized many aspects of life, but its misuse poses a grave threat. Plastics, especially microplastics, enter the marine food web through various sources, potentially impacting plankton, fish, and humans. Despite low current concentrations, the large influx of plastic into the oceans demands concern. While ingested plastics may not cause immediate harm to plankton, toxicity and pollutant accumulation raise alarms. Responsible plastic use, reuse, disposal, and recycling are essential to safeguard our oceans and ecosystems.

Derived from the Greek word plastikos, meaning "capable of being shaped," plastic encompasses a wide range of products with diverse origins and chemical compositions. While some plastics are partially natural, the majority are entirely synthetic. What you may not know is that many plastics are derived from petroleum products, which originated from plankton millions of years ago.

But Is Plastic Truly As Bad As It Is Made Out To Be?

Well, to be honest, plastic itself is not inherently to blame. In fact, it stands as one of the most revolutionary inventions of recent centuries, alongside Sunday afternoon movies and popcorn. The problem lies in our abusive and irrational use of this material. Most plastics are designed for longevity, yet we often treat them as single-use items. To make matters worse, many of these plastics, once discarded, find their way to the sea—a common destination for so many things. In the marine environment, plastics undergo a slow process of degradation, influenced by their chemical composition and environmental conditions. For instance, a plastic bottle can take approximately 600 years to completely degrade (Fig. 19.1). However, since the advent of plastic production in 1860 (a pool ball), numerous plastic containers and materials have started breaking down into minuscule particles called microplastics. These particles, measuring just a few microns (thousandths of a millimeter), are now suspended in varying concentrations throughout the water column or have settled in the sediment. Furthermore, our daily activities generate copious amounts

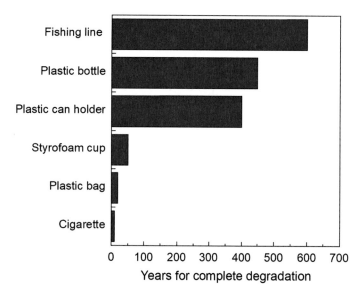

Fig. 19.1 Degradation time of different plastic objects. © Albert Calbet

of plastic fibers and micro-particles. For instance, each time you do a load of laundry (perhaps on Sundays if you are single or nearly every day if you have children), more than 700,000 particles are released, most of which will inevitably find their way into the sea. If you still need more examples, consider that many cosmetics still contain microplastics, the microscopic fragments from tire wear during car usage, and the by-products of industrial activities. These sources contribute to the multitude of origins for marine microplastics. It is worth noting that water treatment plants struggle to efficiently remove microplastics.

Why Should We Be Concerned About Microplastics?

The most pressing issue is that these tiny particles enter the marine food web, either through plankton or via consumption by fish and other organisms. Due to their size, microplastics fall within the range of prey for plankton and fish. Shockingly, approximately 60% of the sardines and anchovies consumed in the NW Mediterranean have plastic particles in their digestive tracts. However, do not despair—the amount of plastic we ingest while entering our brand-new car (mostly comprised of plastic interiors) likely exceeds the plastic content from consuming 1 kg of sardines. In fact, we unintentionally consume around 5 g of plastic per week from various sources, roughly equivalent to a cookie (a plastic one—not so yummy!). Zooplankton encounter similar issues as they sometimes mistake microplastics for prey and ingest them. Laboratory experiments reveal that copepods and protozoa consume plastics when present in high concentrations. Fortunately, the current concentrations of microplastics in the ocean are still relatively low, suggesting that the problem is not yet severe. However, considering that around 8 million tons of plastic enter the sea each year (equivalent to about 500 Eiffel Towers), it becomes evident that the concentration of microplastics in water could pose a genuine threat, even to plankton.

Are Microplastics Harmful?

Generally, ingested plastic particles pass through the digestive tracts of organisms without causing significant harm beyond perhaps a mild constipation. However, certain components used in plastic production, such as plasticizers and other additives, can be toxic. Moreover, certain types of plastic have an affinity for pollutant compounds like hydrocarbons, leading to their accumulation. While the toxicity of microplastics is being extensively studied in laboratory experiments, a comprehensive field approach with more precise analytical techniques is needed to accurately assess the potential hazards.

In conclusion, if you use plastic, strive to reuse it and, when it reaches the end of its useful life, ensure proper disposal and recycling. Collectively, we must work toward preventing our seas from becoming inundated with plastic waste.

20

Why Are There No Insects in the Sea?

© The Author(s), under exclusive license to Springer Nature Switzerland AG 2024
A. Calbet, *The Wonders of Marine Plankton*,
https://doi.org/10.1007/978-3-031-50766-3_20

The absence of insects in marine environments despite their terrestrial abundance is a fascinating puzzle. The challenges of respiratory adaptation and ecological roles, combined with evolutionary ties to plants, provide insights. Insects' deep-sea migration limitations, influenced by their aerial adaptations, contrast with copepods, small crustaceans excelling in marine roles. The ancient origins of copepods further illuminate why insect re-entry into the sea's ecological niche was challenging.

I am sure many of you, who are knowledgeable about this subject, are already thinking "there are insects in the sea." Several species of *Halobates* (Fig. 20.1) indeed inhabit the surface of the ocean, and some other insects can be found in interstitial areas of beaches. However, these occurrences are relatively rare, and insects are generally not found in the depths of the sea. The remarkable contrast between the richness and abundance of

Fig. 20.1 *Halobates* sp. Drawing by Miquel Alcaraz. © M. Alcaraz. With permission

insects on land and their near absence in the sea is a fascinating topic worth exploring. Various theories exist to explain this phenomenon:

Firstly, insects have respiratory systems adapted for aerial environments and cannot exchange gases in water. However, this limitation has been overcome by certain beetles and the larval stages of dragonflies and mosquitoes, which inhabit lakes and rivers. But what happens in the sea? It has been suggested that, being adapted for aerial life, insects could not migrate to the deep areas of the ocean during the day to avoid predation, unlike other similarly sized organisms in the marine realm.

Another factor to consider is the evolutionary history of insects. It is believed that insects evolved from crustaceans over 400 million years ago, and their evolution has been closely intertwined with that of plants. For instance, the appearance of winged insect groups such as butterflies, beetles, and bees coincides with the emergence of flowering plants. Since there are very few flowering plants in the sea, it may explain the absence of many insect groups in marine environments.

Lastly, it is crucial to not only focus on the insects themselves but also consider their ecological roles. Insects fulfill various functions in ecosystems, including herbivory, parasitism, and decomposition. In marine ecosystems, these roles are often performed by copepods, a group of small crustaceans that I have frequently discussed in my blog. Copepods are major herbivores, playing a vital role alongside worms and other organisms in decomposition processes. Additionally, copepods are known to parasitize fish, mollusks, and other marine animals. Their meticulous and efficient execution of ecological functions, coupled with their abundance and biomass, make them unparalleled in the metazoan world.

The evolutionary origins of copepods remain a topic of debate due to the scarcity of copepod fossils. However, recent evidence suggests that they originated during the Cambrian period approximately 500 million years ago. Considering this, it becomes challenging for a group like insects, which transitioned to terrestrial life, to later return to the sea and occupy the ecological niche already established by copepods. Although there have been rare instances of such transitions (e.g., whales, seals, sea turtles, etc.), it is worth pondering whether we should reframe the question posed in the title of this chapter to "Why are there no copepods on land?"

21

Harnessing the Power of Plankton: A Revolutionary Approach to Industry

Recent scientific discoveries have showcased the potential of using plankton for industrial purposes. Plankton offer diverse bioactive compounds for biotechnology, including enzymes and lipids used in biofuels, pharmaceuticals, and aquaculture. Phytoplankton, due to their fast growth and high lipid content, are promising for sustainable biofuel production. Diatoms, with strong silica cell walls, are used in construction for lightweight, durable materials that enhance insulation and indoor air quality. Plankton also offer a solution for global food challenges through nutrient-rich options produced using innovative methods like vertical farming. Challenges include scalability, environmental impact, and economic viability, which require further research and responsible practices.

Recent scientific discoveries have revealed that plankton can also be harnessed for industrial applications, revolutionizing various industries and opening up new possibilities for sustainable and eco-friendly solutions. In this chapter, I will walk you through some of these applications and discuss their pros and cons.

The Use of Plankton in Biotechnology and Other Biology-Related Industries

One of the most promising areas where plankton are being utilized is in the field of biotechnology. Plankton (Fig. 21.1) serve as a rich source of unique and diverse bioactive compounds, including enzymes, lipids, and secondary metabolites, that hold immense potential. For instance, enzymes derived from plankton can be employed in the production of biofuels, pharmaceuticals, and food additives. Lipids extracted from plankton find applications in cosmetics, nutritional supplements (such as omega-3 fatty acids from krill), and aquaculture feed. Plankton-derived secondary metabolites, including certain toxins, have shown promise in the treatment of various conditions such as Alzheimer's disease, cancer, diabetes, AIDS, schizophrenia, inflammation, allergies, osteoporosis, asthma, and pain.

Moreover, plankton have demonstrated great potential in sustainable aquaculture practices, serving as a natural and nutritious food source for

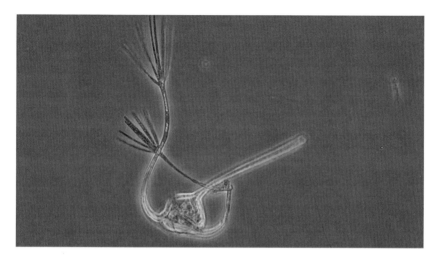

Fig. 21.1 Marine phytoplanktonic dinoflagellate (*Ceratium ranipes*). © Albert Calbet

fish larvae and shrimp. This reduces the reliance on wild-caught fish for fishmeal and minimizes the environmental impact of aquaculture.

The Role of Plankton in Biofuel Production

Biofuels, renewable energy sources derived from biological materials, have emerged as a promising solution to reduce reliance on fossil fuels and mitigate climate change. While traditional biofuels are typically derived from crops such as corn and sugarcane, recent advancements have highlighted the potential of plankton as a sustainable and efficient source of biofuel. Phytoplankton, in particular, offer unique advantages as a biofuel source, thanks to their rapid growth, high lipid content, and potential for sustainable cultivation. The lipids extracted from plankton can be processed to produce biodiesel, offering a renewable and environmentally friendly alternative to traditional fossil fuels. Plankton can be cultivated in controlled environments, such as bioreactors or open ponds, using sunlight, carbon dioxide, and nutrients. Unlike traditional biofuel crop cultivation, plankton can be grown using seawater, reducing the competition for freshwater resources. Additionally, plankton can be cultivated on non-arable land installations, making them a viable option for biofuel production without competing with food crops.

The use of plankton as biofuel presents several additional benefits. Firstly, plankton biofuel is a renewable energy source that can reduce dependence on finite fossil fuels and their contribution to climate change. Plankton biofuel has the potential to significantly reduce greenhouse gas emissions, as the production and combustion of biofuels generally release less carbon dioxide compared to fossil fuels. Since plankton can be cultivated locally, the production of plankton biofuel can be decentralized, reducing the need for long-distance transportation and the associated environmental impacts.

Harnessing Diatoms in Construction

Diatoms (Fig. 21.2), microscopic single-celled algae, not only play a vital role in marine ecosystems but also offer unique properties that make them a promising material for construction. Diatoms possess intricate silica cell walls, called frustules, with diverse shapes and patterns, which

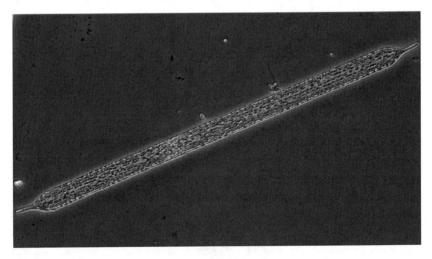

Fig. 21.2 Marine planktonic diatom. © Albert Calbet

can be harvested and utilized for various applications in the construction industry. These frustules exhibit high strength, durability, and thermal stability, making them ideal for various construction applications. Diatom frustules can be harvested from diatomaceous earth, a naturally occurring sedimentary rock composed of fossil diatoms, or they can be cultivated in controlled environments.

One of the primary uses of diatoms in construction is as a sustainable and eco-friendly alternative to conventional construction materials. Diatom frustules can be processed into diatomite, a lightweight and porous material that can be used in the production of cement, concrete, and insulation materials. Diatomite-based construction materials offer several benefits, including improved insulation, reduced weight, and increased durability compared to traditional materials. Furthermore, diatoms have the potential to improve indoor air quality in buildings. This is so, because they have high porosity, allowing for increased air circulation, which can help regulate humidity and absorb indoor air pollutants, contributing to a healthier indoor environment. However, it is important to consider whether the increased air circulation may also result in heat (or cold) escaping from the buildings. The use of diatoms in construction also has the potential to reduce the environmental footprint of buildings. Diatom-based materials have a lower carbon footprint compared to traditional construction materials since diatoms sequester carbon dioxide during their growth process.

A Tiny Solution for Feeding Our Growing Population

I cannot end this chapter without exploring the use of plankton in addressing one of the major challenges faced by human populations: famine. As the global population continues to rise, with estimates reaching 9.7 billion by 2050, finding sustainable and scalable solutions to feed our communities becomes increasingly critical. While traditional agriculture and aquaculture have been the backbone of our food systems, plankton emerges as a tiny yet mighty solution that holds great promise.

One of the main advantages of using plankton as a food source is their remarkable high reproduction rates. Additionally, autotrophic plankton require minimal resources for growth, primarily sunlight and nutrients, making them an efficient and environmentally friendly option for food production. Plankton also pack a nutritional punch. They are rich in essential fatty acids, proteins, vitamins, and minerals, making them a nutrient-dense food source.

Furthermore, plankton can be sustainably harvested using innovative techniques such as vertical farming, where plankton is grown in stacked trays in controlled environments. This allows for year-round production, irrespective of weather conditions, and minimizes the risk of overfishing or depleting natural plankton populations.

The potential of plankton as a food source is not limited to the present but also offers promise for the future. As climate change continues to disrupt traditional agriculture and fisheries, plankton farming can serve as a resilient alternative. Additionally, with its high growth rate and nutrient density, plankton have the potential to address malnutrition and food insecurity, especially in vulnerable populations.

Caveats and Challenges Associated with the Use of Plankton in Industry

Despite the potential benefits, there are challenges and considerations associated with the use of plankton in industry. One of the main challenges is the scalability of their cultivation. While plankton can grow rapidly, achieving large-scale cultivation and production of these organisms is still a technological and logistical challenge that requires further research and development. Another difficulty is the potential environmental impacts of plankton cultivation, such as nutrient pollution, genetic modification, and impacts on marine ecosystems. Careful management, legislation, and monitoring of plankton cultivation practices are necessary to ensure environmental sustainability and minimize potential negative impacts.

Additionally, one of the critical factors to consider is the economic viability and cost competitiveness of plankton production compared to traditional alternatives. The cost-effectiveness of plankton-based solutions is still being evaluated, and further research, technological advancements, and supportive policies are needed to make plankton economically viable and commercially scalable for their various applications.

In conclusion, the harnessing of plankton for industrial purposes represents a revolutionary approach with vast potential. From biotechnology to biofuel production, and from construction materials to addressing food scarcity, plankton offer a wide range of benefits. However, challenges such as scalability, environmental impacts, and economic feasibility must be addressed to fully realize the potential of plankton-based solutions. By continuing research and development efforts, along with responsible management practices, we can unlock the power of plankton and usher in a new era of sustainable and eco-friendly industries.

22

Sampling Plankton

A. Calbet, *The Wonders of Marine Plankton*,
https://doi.org/10.1007/978-3-031-50766-3_22

Plankton, ranging from viruses to large jellyfish, require different sampling methods based on size. For small unicellular plankton like viruses, algae, and protozoa, tools like hydrographic bottles and suction pumps are used to collect samples along with water. Larger zooplankton are captured using nets of varying mesh sizes and towing methods. Preserving samples involves chemical reagents, freezing, or filtration. Sample analysis employs microscopes, specialized machines, and emerging biochemical and molecular techniques. Challenges persist in accurately quantifying species abundance within the complex plankton group.

Plankton encompasses a wide range of organisms varying in size, from viruses as small as 0.00001 mm to large jellyfish measuring up to a meter in length. With this in mind, it is logical to consider that different sampling systems are needed to study plankton based on the size of the target group. In general, we can categorize sampling and capture methods into two groups: those focused on microscopic unicellular organisms and those used for larger organisms, such as multicellular zooplankton.

Sampling Tools for Small Unicellular Plankton

This category includes viruses, bacteria, algae, and protozoa, which are typically collected along with the surrounding water. To accomplish this, we employ various tools, ranging from simple buckets to hydrographic bottles designed to obtain plankton samples at specific depths. A hydrographic bottle is essentially a cylindrical container, often made of PVC or methacrylate, equipped with two airtight lids that can be closed at the desired depth. The most commonly used design is the Niskin bottle (Fig. 22.1b), invented by Shale Niskin in 1966. However, there are other designs like the Nansen and Van Dorn bottles (Fig. 22.1a). These bottles are typically mounted on a structure called a Rosette, which also incorporates devices enabling real-time observation of parameters such as depth, water temperature, salinity, and fluorescence emitted by the algae present. This set of devices is known as a CTD (conductivity, temperature, and depth) and is essential for any oceanographic expedition. Another method for collecting seawater and its accompanying organisms involves

Fig. 22.1 Examples of hydrographic bottles. **(a)** Van Dorn bottles held by the author during an Antarctic Ocean campaign. **(b)** Niskin bottles mounted on a rosette around a CTD. © Albert Calbet

suction pumps that can be deployed to the desired depth. It is important to note that these sampling systems are specifically designed for microscopic plankton. Capturing copepods, fish larvae, or large crustaceans with these methods would yield inaccurate results due to their low abundance and their tendency to escape upon detecting the sampling equipment.

Methods for Capturing Zooplankton

To capture large zooplankton, we use similar methods to the ones used in fishing, although at a much smaller scale. Plankton nets, which typically have a conical shape, vary in size, width, length, and mesh pore diameter, depending on the target group. To provide some context, nets with a mesh pore size of 20 μm (0.02 mm) are suitable for capturing large phytoplankton and microzooplankton, whereas nets with a pore size of

200 μm (0.2 mm) are designed for adult copepods. Nets with a pore size ranging from 0.5 to a few millimeters are used for capturing krill, fish larvae, and other larger organisms. Depending on the target and the type of net, towing can be conducted vertically, horizontally, or obliquely.

The basic structure I have described is the most commonly used and is associated with nets such as the Juday Bogorov, WP2 (Fig. 22.2a), and Bongo (Fig. 22.2b), among others. However, there are more complex and mechanized nets available, such as the Bioness or LHPR (Longhurst Hardy Plankton Recorder; Fig. 20.2c). These advanced nets enable sampling at different depths and simultaneously capture specimens with various mesh sizes. They also provide a record of water depth and physicochemical parameters. Some devices even incorporate video cameras that capture images or recordings of everything within their field of view.

Certain plankton organisms, particularly gelatinous plankton, are delicate and challenging to capture intact using plankton nets. To obtain living and pristine specimens of specific gelatinous plankton species, we

Fig. 22.2 Examples of fishing nets—(a) Double WP2 net. (b) Bongo nets. (c) LHPR. © Albert Calbet

have no choice but to immerse ourselves and carefully extract them one by one from their environment.

Preservation of Samples

Now that we have collected our plankton samples, the next step is to decide how to process them. Similar to the capturing devices, the processing methods depend on the specific group being studied. Generally, samples can be examined live or preserved for future analysis. Preservation can be achieved using chemical reagents like formaldehyde or Lugol's solution (a modified iodine tincture). Alternatively, samples can be frozen at temperatures of -20 or -80 °C, depending on the desired analysis. Filtration and drying, or filtration and extraction of pigments in solvents like acetone or ethanol, are also common preservation techniques. In some cases, a combination of techniques may be necessary, such as fixing, filtering the sample, and subsequently freezing it. This last technique is employed when observing unicellular organisms using an epifluorescence microscope.

Sample Analysis

To observe and classify plankton, the most practical approach is to use stereoscopic magnifiers or microscopes. However, specialized machines allow for faster sample processing, albeit with reduced taxonomic resolution compared to an experienced human specialist. For instance, flow cytometers are utilized for bacterial analysis, the FlowCam is employed for algae and microzooplankton, and the Zooscan is used for larger zooplankton. These devices operate on the same principle. The sample passes through a tube or is placed on a plate, where it is stimulated by a light source (such as a laser or normal light), and the emitted photons or captured images are recorded by specialized video or photo cameras. Complex computer programs often process these images or light emissions to estimate abundance and perform initial classification of the organisms present in the water.

Recently, biochemical and molecular analysis techniques for plankton samples have gained popularity. These emerging techniques provide insights into the diversity and certain physiological processes occurring in the water. However, accurately quantifying the abundance of all species within the complex plankton group is still a significant challenge.

23

Myths and Legends of Modern Research: Experimenting with Zooplankton

As a zooplankton-focused marine biologist, I often clarify that my work involves understanding the role of these small organisms in marine ecosystems. I study their diet, impact, and responses to environmental factors. Research projects, funded through competitive applications, drive our investigations. We either gather samples during oceanographic cruises or cultivate organisms in controlled labs. Collaboration is integral to our field, involving joint campaigns, data sharing, and co-authored research papers. While not as glamorous as coral reefs, whales, or shark research, our work contributes to a deeper understanding of the marine world and its preservation.

Often, when someone asks me about my profession, I say, "I am a marine biologist." Immediately, upon hearing these words, the mind of my interlocutor drifts to dives in coral reefs in paradise-like places, shark research, or observing cetaceans in the open sea. Then, I feel compelled to clarify by saying that I research plankton, more specifically zooplankton. The next step is a smirk, a mix of disappointment and confusion, followed by the question: What are zooplankton? Tired of the same old speech, I give a somewhat imprecise generalization: they are what whales eat (krill).

What Does a Marine Biologist, Dedicated to the Study of Zooplankton, Actually Do?

Basically, we try to understand the function of this complex group of organisms in the marine ecosystem. In particular, I study their role in planktonic food webs. In simpler terms, I study what they eat, their impact, the environmental variables that affect their behavior and physiology, etc. I assure you, there is no shortage of work. I will now explain what we do, in broad strokes, to give you a glimpse of the profession I have chosen, or that has chosen me; who knows?

Research Projects

Our research is based on research projects; therefore, our success relies on our ability to attract funding, whether national or foreign, to carry out our investigations. This means that every time we come up with an interesting question, we have to wait for a project call (usually once a year) and then apply for it. With luck, our project, evaluated by anonymous reviewers, gets funded (the success rate varies widely from country to country: in Spain it is around 50%, much lower, 13%, for EU proposals, and for USA NSF proposals it is approximately 25%). An average research project lasts for 3 or 4 years. Once we have the funding, we should start the research. In our case, that means we have to figure out how to obtain the experimental organisms. We can do that by going to the sea to catch them or by establishing laboratory cultures. I will now explain the pros and cons of each approach.

Field Sampling, from Getting Our Feet Wet on the Beach to Major Oceanographic Campaigns

As marine biologists, we consider the seas and oceans as our study and inspiration source. That is why it is common for us to take field samples from time to time. Depending on the goal of our study, we can simply dip our feet in the beach and fill some containers with water, take a small boat and sample coastal or offshore waters, or participate in a major oceanographic cruise. The latter option is often the most appealing from the outside. After all, sailing the ocean has always been deeply rooted in our dreams. We all have a little explorer vein that often breaks when we are fully aware, in the middle of a storm crossing the Drake Passage with 6-m waves, that we get seasick. In those moments, when you struggle to keep into your stomach the last bite of salami sandwich and hold onto any object fixed to the floor or wall, that is when you truly realize that this is your vocation. That being said, I must confess that navigating the waters of the Gerlache Strait in the Antarctica (Fig. 23.1) is among the most captivating and impressive spectacles you can ever see.

But we have not come here to enjoy the view; we have come to work. I confess that the work on an oceanographic cruise is tough. We are often busy for long hours without rest, tired, worn out by the routine that repeats relentlessly every day, enduring cold or heat depending on the climate where we sample, feeling a bit seasick... well, you get the idea. Then there is the food. Usually, with a few exceptions, ship food is not like grandma's cooking.

Fig. 23.1 The Gerlache Strait on a sunny day (Antarctica). © Albert Calbet

In any case, as a zooplankton specialist, I usually divide my activities on the ship between laboratory work and deck sampling with different fishing gears (nets adapted for catching plankton), collecting water samples with hydrographic bottles, or assisting in common sampling tasks. In the laboratory, that is when the delicate work begins. With the help of a stereomicroscope and great care not to damage them, we separate the most abundant groups of zooplankton by species and gender, introduce them alive, very carefully, into bottles of different volumes, and take the necessary measurements for the study. It seems easy, but the constant movement of the ship, the fatigue, and the very limited space, which are common in oceanographic campaigns, continually hinder our task.

Experimentation with Laboratory Cultures

The alternative to oceanographic campaigns and other field sampling is working with laboratory organisms. When working with cultured organisms, we usually play it safe. We know what we have, and we do not depend on what is in the sea at the time of year we are in. However, the majority of plankton organisms have not been successfully cultured so far. Besides being able to work with only a few species, maintaining them in culture for long periods requires a lot of effort, luck, and patience. Unfortunately, accidents are common, and for some reason, often the cultures collapse and die. Keep in mind that many species require almost daily attention. Nonetheless, despite the challenges, working with laboratory cultures offers numerous advantages. It allows us to control the environmental conditions, such as temperature (Fig. 23.2), salinity, and light intensity, providing a stable and reproducible experimental setup. We can also manipulate the food availability and composition, allowing us to study the effects of different diets on the growth and development of the organisms. Additionally, laboratory cultures provide the opportunity for long-term observations and experiments that would be impractical or impossible to conduct in the field.

Working with laboratory cultures involves maintaining a dedicated facility with the necessary equipment and infrastructure. It requires expertise in culturing techniques, such as providing suitable culture media, ensuring proper aeration, and maintaining cleanliness to prevent

Fig. 23.2 Laboratory set-up for growing copepods at different temperatures. Marine Sciences Institute, CSIC. Barcelona. Spain. © Albert Calbet

contamination. Monitoring the health and growth of the organisms is essential, and regular observations and measurements are conducted to track their development and reproductive cycles.

In addition to studying the basic biology and life cycle of the organisms, laboratory cultures can be used for a wide range of experiments. For example, we can investigate the effects of environmental stressors, such as temperature changes or pollution, on the physiology and behavior of the organisms. We can also study the interactions between different species or examine the effects of specific factors on their population dynamics. These controlled experiments (Fig. 23.3) allow us to gain valuable insights into the responses and adaptations of planktonic organisms to environmental changes.

Fig. 23.3 Example of a basic experimental set-up with marine protozoans. © Albert Calbet

Collaborative Research and Data Sharing

Marine biology is a highly collaborative field, and researchers often work together to address complex scientific questions and share resources and data. Collaboration can take various forms, from joint field campaigns and laboratory experiments to data sharing and co-authorship on research publications. Collaborative field campaigns bring together experts from different disciplines and institutions, combining their expertise and resources to tackle larger-scale projects. These campaigns often involve multiple research vessels, each with its specialized equipment and sampling protocols. By pooling their resources and knowledge, scientists can obtain a more comprehensive understanding of the marine environment and its inhabitants.

In addition to collaborative fieldwork, researchers also engage in data sharing and integration. This is particularly important in the era of Big Data, where vast amounts of information are generated from various sources, including remote sensing, satellite observations, and sensor networks. By sharing data, researchers can combine datasets from different regions and periods, allowing for broader-scale analyses and more robust conclusions. Open data initiatives and online repositories play a crucial role in facilitating data sharing and accessibility for the scientific community.

Furthermore, collaboration extends to the dissemination of research findings. Scientists often collaborate on research papers, where they contribute their expertise and data to collectively produce a comprehensive study. Co-authorship allows researchers to share credit and recognition for their contributions to the project. Collaboration also extends beyond individual research groups to international collaborations, where scientists from different countries work together to address global marine issues and conservation challenges.

In conclusion, marine biology encompasses a broad range of research activities, from fieldwork and laboratory experiments to collaborative efforts and data sharing. Whether studying the intricate ecosystems of coral reefs, unraveling the mysteries of deep-sea life, or investigating the dynamics of planktonic organisms, marine biologists are dedicated to advancing our understanding of the oceans and their inhabitants. Through their efforts, they contribute to the conservation and sustainable management of marine resources, striving to protect and preserve these invaluable ecosystems for future generations.

24

A Brief History of Plankton Discovery

A. Calbet, *The Wonders of Marine Plankton*,
https://doi.org/10.1007/978-3-031-50766-3_24

In this chapter, I provide you with a glimpse into the early interactions between humans and plankton, as well as how we managed to study these fascinating creatures. As you will see, there is evidence dating back to ancient Greece, which suggests the existence of strange beings, including plankton organisms, that did not fit into any known classification. However, a significant amount of work was done by a series of amateur and professional naturalists from the sixteenth to the nineteenth centuries. While their discoveries were not always published as books, they were often communicated through correspondence with renowned institutions like the Royal Society of London. Although many valuable records are lost, a substantial portion of this correspondence is still available, providing evidence of the exciting progress made by these individuals as they ventured into a whole new world of discovery. In this regard, I will particularly emphasize the significance of the first illustrations of plankton, as the popular saying goes, "a picture is worth a thousand words."

Multicellular Plankton

Logically, the earliest recorded members of plankton were likely jellyfish and other large organisms, especially those that lived in association with other animals for daily consumption, such as fish. In the fourth century BC, the great philosopher and naturalist Aristotle first identified a parasitic copepod. He classified it, along with jellyfish and other soft-bodied organisms like sponges and ascidians, under the name Zoophyta, a category that lasted for hundreds of years. However, credit for the first image of a parasitic copepod (Fig. 24.1) goes to Gillaume Rondelet, a French zoologist born in 1507.

The discovery of free-living copepods had to wait a little longer. In 1688, Stephan Blankaart drew the first recorded illustration of a free-living copepod (Fig. 24.2) from a freshwater sample. Carl Linnaeus, the father of modern taxonomy (1707–1778), named them *Monoculus* and classified them as insects. This classification remained in place for many years until the early nineteenth century when Jean-Baptiste Lamarck classified them as crustaceans, along with water fleas, amphipods, and isopods.

Interestingly, the first illustration of a free-living marine copepod corresponds to Gunnerus (1770), a Norwegian zoologist who identified

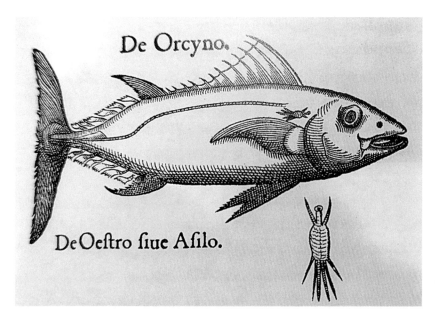

Fig. 24.1 In the illustration by Rondelet (1554), you can see what appears to be a parasitic copepod below, to the right, and above the fish gills

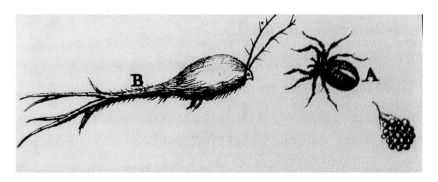

Fig. 24.2 Appendix 24.2. The first illustration of a free-living copepod, possibly of the genus Cyclops, from a freshwater sample. Stephan Blankaart (1688)

Calanus finmarchicus, a species of significant abundance and importance on the Norwegian coast, although he referred to it as *Monoculus finmarchicus* (Fig. 24.3).

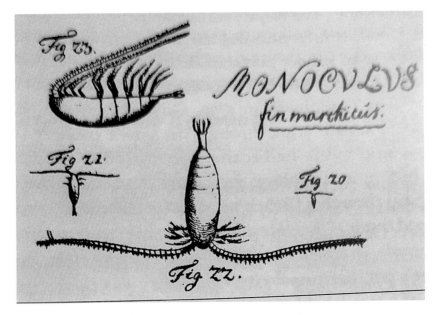

Fig. 24.3 The first illustration of a free-living marine copepod. Ernst Gunnerus (1770)

Unicellular Plankton

The discovery of protists, the unicellular plankton, is credited to Antonie van Leeuwenhoek. Using his rudimentary microscope, the Dutchman, recognized as the father of microbiology, observed infusoria and other planktonic organisms. Between 1674 and 1716, he described several species of protozoa, primarily ciliates (infusoria, Fig. 24.4), among other planktonic creatures.

He did not pay much attention to planktonic algae, and although he likely observed them, the first description of a diatom (not strictly planktonic) was made by an English gentleman, probably Charles King, in 1703 in a note sent to the Royal Society of London. Subsequently, many naturalists dedicated themselves to classifying, observing, and illustrating protozoa. For example, O.F. Müller provided a detailed description with drawings of the behavior of tintinnids in 1779. However, the most renowned artist and scientist depicting these beautiful creatures was the

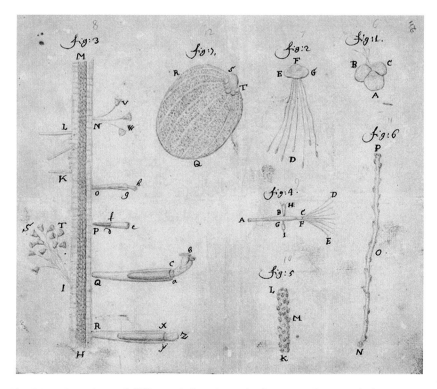

Fig. 24.4 Drawings of different infusoria and other organisms made by Antonie van Leeuwenhoek in 1702

German Ernst Haeckel (Fig. 24.5). In 1866, Haeckel suggested that all the animals (including bacteria) he observed under the microscope should form a third independent animal kingdom called Protista, which was the first and primordial kingdom.

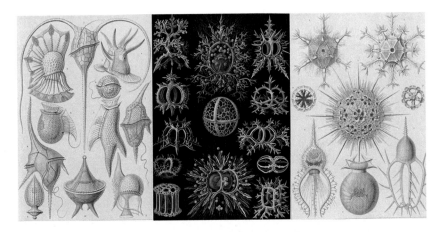

Fig. 24.5 Illustrations of different protozoa by Ernst Haeckel

Plankton Research

Many years passed from the discovery and classification of planktonic organisms to the beginning of research on their role in the oceans. The initial studies, which primarily focused on descriptive observations of life forms, were linked to large expeditions such as Captain James C. Cook's voyages between 1768 and 1780 in the Pacific Ocean. These expeditions provided evidence of phenomena like spots of the cyanobacterium *Trichodesmium* on the surface of the sea. However, the first nets specifically designed to collect plankton were likely used by French naturalists François Péron and Charles-Alexandre Lesueur during their expedition to Australia from 1801 to 1804. Similarly, Charles Darwin utilized nets to collect plankton samples during the extended voyage of the HMS Beagle from 1831 to 1836.

Nevertheless, it was the Challenger expedition (1872 to 1876) that played a pivotal role in the development of modern oceanography. Equipped with laboratories, microscopes, and a team of six scientists, the HMS Challenger was the first expedition specifically organized to gather data from the marine environment. These data included temperature, water chemistry, bottom geology, currents, and marine life, thereby laying the foundation for comprehensive marine research.

In 1887, physiologist Victor Hensen coined the term "plankton" to describe the animals drifting in water currents. He was also the first to propose the idea that marine life might be sustained not by the influx of rivers but by microscopic primary producers. Through his studies, Hensen improved upon existing designs of plankton nets and developed a quantitative net that is still used today—the Hensen net. However, based on his samples, he mistakenly concluded that plankton were uniformly distributed throughout the ocean and that there were insufficient numbers to support fisheries. These conclusions were soon proven incorrect.

Research on the distribution and abundance of different plankton groups continued throughout the late nineteenth and early twentieth centuries, led by scientists such as Marie Lebour (a specialist in diatoms and dinoflagellates, 1876–1971), Alister Hardy (creator of the continuous plankton capture system, CPR, 1896–1985), Sheina Marshall (a pioneer in copepod feeding studies, particularly *Calanus* sp., 1896–1977), and Hans Utermöhl (inventor of the sedimentation chambers named after him, 1896–1984). These studies resulted in significant discoveries, including the identification of daily patterns of vertical migration. The French naturalist George Cuvier observed the vertical migration of zooplankton in 1817, although his study was limited to a single day and conducted in a lake. A more comprehensive and enduring investigation into vertical migration was conducted by the German scientist Carl Chun in 1888.

Despite hints about their behavior, much of what these microscopic organisms actually did in the sea remained a mystery until the twentieth century. It was during this time that estimates of primary production and zooplankton production began to emerge. Initially, variations in oxygen measurements under light and dark conditions were used to estimate primary production. Later on, radioactive carbon became a valuable tool for these estimations. Similarly, the estimation of zooplankton production as a technique developed well into the twentieth century.

In 1963, Ramon Margalef laid the foundation for understanding ecosystem structure and the importance of ecosystem maturity in species succession. Toward the end of the twentieth century, scientists such as

F. Azam, T. Fenchel, and others introduced the concept of the microbial loop, emphasizing the vital role of bacteria and dissolved organic compounds in the marine food web.

These milestones have been crucial in advancing our understanding of plankton, but we are still in the early stages of comprehending and predicting the functions of different plankton groups in the marine food web. Despite the remarkable progress we have made and the availability of modern techniques and disciplines like satellite imaging, genomics, and gene expression, there is much more to uncover in the fascinating world of plankton.

Glossary

Alga Eukaryotic organism capable of performing oxygenic photosynthesis. In marine environments, dominant algae are unicellular. Plural algae.

Algal bloom A rapid and extensive growth of algae in aquatic environments, often occurring in temperate seas during early spring.

Appendicularia A small tunicate featuring a tail with a notochord used for propulsion and creating feeding currents. These organisms inhabit gelatinous filters formed by their own secretions.

Autotroph An organism that synthesizes complex organic compounds from carbon dioxide. Terrestrial plants and marine algae are examples of autotrophs, also known as primary producers.

Bioluminescence The emission of light by a living organism through a chemical reaction.

Ciliate A unicellular organism distinguished by the presence of cilia, which it employs for movement or feeding.

Cladoceran A diminut freshwater or marine crustacean, also referred to as a water flea.

CO_2 Abbreviation for carbon dioxide, a by-product of respiration and the combustion of fossil fuels. It plays a significant role in global warming and the greenhouse effect.

A. Calbet, *The Wonders of Marine Plankton*,
https://doi.org/10.1007/978-3-031-50766-3

Copepod A minute crustacean of paramount importance in marine food webs.

Cyanobacteria A group of bacteria capable of photosynthesis.

Diatom A unicellular algae species possessing a siliceous skeleton.

Dinoflagellate A unicellular organism equipped with two flagella.

Doliolid A free-living tunicate belonging to the gelatinous plankton, character-ized by its barrel-like shape. It sustains itself by filtering particles from the water.

Epifluorescence microscopy A microscopy technique that employs light at vary-ing wavelengths to induce fluorescence in the sample.

Eukaryotic An organism with cells containing a membrane-bound nucleus and other organelles. This includes plants, animals, fungi, and protists, distin-guishing them from prokaryotes.

Flagellate A unicellular organism propelled by one or more flagella.

Food web An intricate network of feeding interactions that define an ecosystem. Unlike the classical concept of a linear food chain in the sea, recent research has revealed interconnected food chains.

Geologic time scales A chronological framework encompassing Earth's history, often spanning billions of years in the context of this book.

Heterotroph An organism reliant on consuming organic matter for survival.

Holoplankton Organisms that complete their entire life cycle within the plank-ton community.

Ichthyoplankton The eggs and larvae of fish.

Krill Euphausiids, small shrimp-like crustaceans that serve as a primary food source for many whale species.

Marine snow Aggregates of particles that occur in the deep ocean.

Meroplankton Organisms that spend part of their life cycle as plankton and the rest outside of it.

Mesozooplankton The zooplankton subset ranging in size from 0.2 to 20 mm.

Microzooplankton The zooplankton subset ranging in size from 20 to 200 μm.

Mixotroph An organism combining autotrophic and heterotrophic feeding strategies.

Nauplius The larval stage of a crustacean.

Photic zone The water column layer in marine and freshwater ecosystems where sunlight can penetrate.

Photosynthesis The process by which plants (including phytoplankton) use sun-light, water, and carbon dioxide to produce oxygen and energy in the form of sugar.

Phytoplankton The plant component of the plankton community.

Plankton   A collection of mainly small organisms that drift in the water of seas, lakes, and rivers.

Pluricellular   An organism composed of multiple cells.

Primary production   The creation of new organic matter by phytoplankton through photosynthesis, originating from inorganic substrates.

Prokaryote   A type of cellular organism that lacks a distinct membrane-bound nucleus and other membrane-bound organelles. Prokaryotes are typically unicellular and belong to the domains Bacteria and Archaea.

Protist   A eukaryotic unicellular organism of either animal or plant origin.

Protozoan   A eukaryotic unicellular organism of animal or heterotrophic nature.

Red tide   The accumulation of unicellular algae, toxic or non-toxic, in a specific area, varying in extent and duration. In scientific terms, "harmful algal bloom" is the preferred term.

Salp   A tunicate belonging to the gelatinous plankton, resembling a bag or open-ended tube. It captures particles from the water through filtration. While free-living, it can also form colonies or lengthy chains.

Suspension feeder   An organism that consumes suspended particles.

Thermocline   A water column layer that demarcates zones with different temperatures.

Tunicate   An animal subphylum encompassing ascidians, salps, doliolids, and appendicularians. The term "tunicate" alludes to their secretion of a tunic made of cellulose substance (tunicin); the larvae of these organisms possess a notochord in the tail region.

Unicellular   Comprising a single cell.

Zooplankton   The animal constituents of the plankton community.

Supplementary Bibliography[1]

Calbet A (2008) The trophic roles of microzooplankton in marine systems. ICES J. Mar. Sci. 65:325–331

Calbet A (2022) El plancton y las redes tróficas marinas. Colección "¿Qué sabemos de?". Editorial CSIC y Catarata, p 104 páginas. ISBN 978-84-00-11087-1

Calbet A, Garrido S, Saiz E, Alcaraz M, Duarte CM (2001) Annual zooplankton succession in coastal NW Mediterranean waters: the importance of the smaller size fractions. J. Plankton Res. 23:319–331

Castellani C, Edwards M (2017) Marine plankton. A practical guide to ecology, methodology, and taxonomy. Oxford University Press, Oxford

Margalef R (1978) Life-forms of phytoplankton as survival alternatives in an unstable environment. Oceanologica Acta 1:493–509

Miller CB, Wheeler PA (2012) Biological oceanography, 2nd edn. Wiley-Blackwell, New Jersey, U.S.A. 464 pp

Sardet CE (2015) Plankton: Wonders of the drifting world. The University of Chicago Press, Chicago, IL, U.S.A. 222 pp

Schmoker C, Hernández-León S, Calbet A (2013) Microzooplankton grazing in the oceans: Impacts, data variability, knowledge gaps and future directions. J. Plankton Res. 35:691–706

---

[1] This book is not intended to serve as a reference or as textbook. Its objective is to pique curiosity and educate the reader about the significance, intricacy, and vulnerability of plankton. Consequently, I have refrained from incorporating scientific references to support the presented arguments. Nevertheless, for those inclined to explore the intricacies of plankton through alternative scientific and outreach publications, I present a comprehensive list of scholarly papers and informative books that delve into this subject matter.

© The Author(s), under exclusive license to Springer Nature Switzerland AG 2024
A. Calbet, *The Wonders of Marine Plankton*,
https://doi.org/10.1007/978-3-031-50766-3